BIOETHICS

Donna Dickenson

ALL THAT MATTERS

ALL THAT MATTERS

Hodder Education

338 Euston Road, London NW1 3BH.

Hodder Education is an Hachette UK company

First published in UK 2012 by Hodder Education

First published in US 2012 by The McGraw-Hill Companies, Inc.

Copyright © 2012 Donna Dickenson

The moral rights of the author have been asserted

Database right Hodder Education (makers)

British Library Cataloguing in Publication Data: a catalogue record for this title is available from the British Library.

Library of Congress Catalog Card Number: on file.

10 9 8 7 6 5 4 3 2

The publisher has used its best endeavours to ensure that any website addresses referred to in this book are correct and active at the time of going to press. However, the publisher and the author have no responsibility for the websites and can make no guarantee that a site will remain live or that the content will remain relevant, decent or appropriate.

The publisher has made every effort to mark as such all words which it believes to be trademarks. The publisher should also like to make it clear that the presence of a word in the book, whether marked or unmarked, in no way affects its legal status as a trademark.

Every reasonable effort has been made by the publisher to trace the copyright holders of material in this book. Any errors or omissions should be notified in writing to the publisher, who will endeavour to rectify the situation for any reprints and future editions.

Hachette UK's policy is to use papers that are natural, renewable and recyclable products and made from wood grown in sustainable forests. The logging and manufacturing processes are expected to conform to the environmental regulations of the country of origin.

www.hoddereducation.co.uk

Typeset by Cenveo Publisher Services.

Printed in Great Britain by CPI Group (UK) Ltd, Croydon, CR0 4YY

Also available in ebook

Contents

Acknowledgements

The author and publishers would like to thank the following for their permission to reproduce photos in this book. Chapter 1: © Argus – Fotolia (page 3); © Georgios Kollidas – Fotolia (page 4); © SuperStock (page 6); public domain/http://commons.wikimedia.org/wiki/File:Frankenstein.1831.inside-cover.jpg (page 7); © Moviestore Collection / Rex Features (page 11); Mary Evans Picture Library (page 15). Chapter 2: © Moviestore Collection / Rex Features (page 25); public domain/http://commons.wikimedia.org/wiki/File:PSM_V03_D380_John_Stuart_Mill.jpg (page 26, top); © Alexandr Mitiuc – Fotolia (page 26, bottom); © Ajit Solanki/AP/Press Association Images (page 32). Chapter 3: © www.CartoonStock.com (page 40); © ProMotion – Fotolia (page 52). Chapter 4: Library of Congress Prints & Photographs division /LC-USZC4-4311 (page 59); Chapter 6: © Ulf Andersen/Getty Images (page 70); © Chung Sung-Jun/Getty Images (page 88).

1

Should we do whatever science lets us do?

Human biotechnologies . . . promise, and sometimes deliver, results that in ancient times were thought to be achievable only by divine intervention: they heal the sick, let the crippled walk again, and give children to the barren. They act directly on our bodies, behaviors, and minds and alter the way we understand ourselves in the process.[1]

ALL THAT MATTERS

New biotechnology is the theatre of both our greatest hopes and some of our greatest fears. The thinkers of the eighteenth-century Enlightenment may have believed that scientific progress would be accompanied by greater social justice and happiness – but what if scientific advances are used to increase social injustice rather than diminish it? A well-documented Canadian report, for example, criticized the misuse of transplant technology to provide kidneys and other organs from executed Chinese prisoners for wealthy Westerners.[2]

This, and other instances of 'body shopping',[3] such as the international market in human eggs and the development of an international commercial surrogacy hub in India,[4] would have been impossible and even unforeseeable before medical advances in assisted reproduction, genetics, immunosuppressive drugs and molecular biology. Typically these medical advances are themselves genuinely beneficial: I want to make that clear from the start, because anyone who criticizes the consequences of these developments is often wrongly and cheaply labelled a knee-jerk opponent of progress. But the uses to which medical advances are put can also be misuses. This necessary reality – sometimes denied by the more extreme advocates of new technologies – has generated bitter debates, which sometimes surpass even the attacks on Darwin's theory of evolution in their ferocious implacability.

The role of bioethics – from the Greek words bios (life) and ethos (values or morality) – is to debate those issues more rationally, to make sure that the onward march of science doesn't trample down vulnerable populations, to prevent harms from outweighing benefits, to ask whose interests prevail and to raise questions about whether justice is being served by new scientific developments. That's how I see my own role as a bioethicist.

Justice

Justice was defined by the Greek philosopher Aristotle as treating equals equally and unequals unequally. The contentious questions are about who counts as equal. In Aristotle's own time slaves and women did not, and Aristotle – unlike his tutor Plato – was very much a man of his time in accepting those social conventions. We no longer accept such inequalities as natural or inevitable – at least not overtly – but Aristotle's formulation 'treating equals equally' is still important in bioethics, because it expands the horizon of our considerations to ensure that new scientific developments do not harm vulnerable groups disproportionately.

Important though I think all those issues are, however, there is also an even more profound question –

▲ The Greek philosopher Aristotle defined justice as 'treating equals equally and unequals unequally' – but who counts as equal?

metaphysical as well as moral. If our life in the state of nature is 'solitary, poor, nasty, brutish and short', as the seventeenth-century political theorist Thomas Hobbes wrote, then overcoming the limitations imposed on us by nature is itself our greatest hope. To some, that could rightfully include enhancing our cognitive capacities to the extent that we also alter human nature itself.

However, others think that the demise of nature is the eventuality we should fear most. As this quotation from the British sociologist Sarah Franklin asserts, if science can remove any natural limitations on our ambitions, then we must voluntarily accept the responsibility of imposing

▲ Thomas Hobbes, the seventeenth-century author who called life in the state of nature 'solitary, poor, nasty, brutish and short'.

those limits on the devices and desires of our own hearts, which may as easily be pernicious as beneficial:

> *We can no longer assume that the biological 'itself' will impose limits on human ambitions. Humans must accept much greater responsibility towards the realm of the biological, which has become a wholly contingent condition.*[5]

Many others, however, invest the new biotechnology with all the aspirations and faith once accorded to religious salvation. If biotechnology's most fervent acolytes are right, whether or not there are inevitable boundaries on what science can achieve, there should be no limits imposed on its progress by ethical questioning, political interference or legal prohibitions. If the realm of the biological could ever become 'a wholly contingent condition', that would be an entirely good thing, in this alternative view. We tamper with biomedical science at our peril. If we want to enjoy the fruits of scientists' labours – and it would be hypocritical of us to say that we don't – then our job is to help science overcome disease, pain and early death: our natural condition.

Yes, in this view, we must indeed accept greater responsibility towards the realm of the biological, but not in the sense of critique or prohibition: quite the reverse. The task of bioethics, such analysts would say, is merely to act as an intelligent advocate for science, debunking myths and allaying irrational fears – and, almost by definition, all fears are irrational when compared with the immense benefits of modern science. Although philosophers may have seen themselves as gadflies since the death of Socrates at the hands of the Athenian state, that's no longer

▲ Jacques-Louis David, 'The death of Socrates': executed by the Athenian state, supposedly for dissidence. Should bioethicists be critics like Socrates?

their role – or, at least, it's not the role of philosophical bioethics, these advocates would say. The critical skills of philosophers remain central, but their role, in this view, is to overcome popular misapprehensions.

One strong supporter of something like this approach, the British bioethicist John Harris, goes one step further. Not only do we have a passive or negative responsibility not to interfere with science in its production of better living conditions, he says, we actually have an active, positive duty to volunteer as research participants:

> 'Frankenstein science' is a phrase never far from the lips of those who take exception to some aspect of science or indeed some supposed abuse by scientists. We should not, however, forget the

powerful obligation there is to undertake, support, and participate in scientific research, particularly biomedical research, and the powerful moral imperative that underpins those obligations.[6]

Because we all benefit significantly from modern medicine, this argument runs, we are all obliged to do our part in advancing the state of medical knowledge. Harris thinks that this truth is self-evident to any right-

▲ 'Frankenstein science': an illustration by Theodor von Holst from the frontispiece of the 1831 edition of Mary Shelley's novel.

thinking person. As he adds, 'The obligation to participate in research should be compelling for anyone who believes there is a moral obligation to help others, and/or a moral obligation to be just and do one's share. Little can be said to those whose morality is so impoverished that they do not accept either of these two obligations.'[7]

That's certainly claiming the moral high ground. But do we all benefit equally from scientific progress, and are we all called on to make equal sacrifices? As always in bioethics, we need to ask whose interests are being served. Individuals aren't the only beneficiaries of medical advances: modern biomedicine is highly commercialized, globalized and, in many cases, massively profitable.[8] The US commentator Gina Maranto thinks that there is a potentially exploitative imbalance between the starry-eyed motivations of research subjects and the hard-nosed realities of modern science: 'Given the market conditions under which science operates, the freely given gift quickly becomes commoditized.'[9] And one-way altruism is better known as exploitation.

We don't all benefit equally in another sense, either: increasingly, research is either outsourced to developing countries or conducted by private contract research organizations. In 2004, for example, drug companies launched over 1,600 overseas trials, while more than 70 per cent of domestic trials were conducted by the profit-making sector. The US research subjects are often long-term unemployed people or undocumented immigrants, participating in one trial after another as their main source of income.[10] Although Third World populations may derive temporary benefits from participation, such as health care for the duration of the trial, they won't necessarily enjoy the long-term benefits of the new drug. Neither will the long-term unemployed, if they have no health coverage.

As another US bioethicist, Stuart Rennie, has written in a recent critique of Harris's claim:

> It is probably no accident that calls to change the moral status of research participation [from voluntary to morally obligatory] come at a time when health research is increasingly being outsourced and offshored around the world: relocating research in low-income countries and framing research participation as a robust duty are mutually reinforcing ways of enhancing power and profitability. So at least in the short term, the chief beneficiaries of calling research participation 'obligatory' are researchers, research institutions, public and private research funding agencies, and pharmaceutical companies.[11]

The supposed 'duty' to participate in research trials was spectacularly discredited in the case of Hwang Woo Suk, a South Korean stem cell researcher who claimed in 2005 to have created 11 cell lines that could eventually yield personalized 'spare parts kits'. Hwang's particular stem cell technique, somatic cell nuclear transfer (SCNT), requires huge numbers of human oocytes (egg cells). When he made his announcement – later proved to be false – other researchers were agog to know where he'd obtained the necessary numbers of oocytes. The message they got was that Korean women had dutifully donated their eggs for the greater national good.

But as the Canadian bioethicist Francoise Baylis[12] has shown, that was a subterfuge. A majority of Hwang's 'donors' (75 per cent) were either paid or given treatment in kind. (Payment is usually seen in research ethics as an unfair inducement, which might persuade the seller to take more risks than is sensible.) One of Hwang's doctoral students later wrote that, although she had claimed that she was motivated by her sympathy for sick children and her patriotic feelings for South Korea, she was actually put under pressure to 'donate' if she wanted her career to flourish.

The Korean Ministry of Health reports that Hwang used a total of 2,221 eggs from 119 women – all to no avail. Of 79 women who sold their eggs through one hospital, 15 were treated for ovarian hyperstimulation, which can be fatal. Two women were hospitalized three times for the condition; oocytes were taken a second time from one of them, despite the harm she had already suffered. Other women were already undergoing fertility treatment, which they received at a lower price on condition that they donated some of their eggs (the practice known euphemistically as egg 'sharing'). But their better eggs were taken for Hwang's research, lessening their own chances of successful pregnancy.

Of course it could be countered that Hwang was an atypically dishonest researcher and that his informed consent procedures were blatantly unsatisfactory. Nor would it be fair to imply that it was necessarily wrong to undertake the research because it turned out unsuccessfully. But the bottom line remains that, in this type of stem cell research, as in infertility research and other areas of reproductive medicine, it is women's tissue that is required, and it is women who will bear a disproportionate share of the burdens of altruistic participation. We don't all benefit or sacrifice equally in aiding scientific progress.

I think we also need to be cautious about the idea that those who identify abuses of biomedicine are prone to accuse researchers of 'Frankenstein science'. One could just as well say that those who think that bioscience can do no wrong are falling into the trap of relying on a cliché rather than empirical evidence. It's a way of dismissing critics who point to possible genuine harms as scaremongers

▲ Another pessimistic view of the state of nature: a still from the film *Lord of the Flies*.

and depicting the general public as gullible ignoramuses, while claiming the rational high ground for science.

In an article called 'Questioning the sci-fi alibi',[13] the British social scientist Jenny Kitzinger presents new evidence that disproves the stereotype of a scientifically ignorant and emotional public. In 43 interviews and 20 focus group discussions about new biotechnologies such as stem cells or cloning, Kitzinger discovered that, far from retailing nightmares from *Jurassic Park*, her subjects were actually very apologetic about introducing any fears that might be dismissed as 'science fiction'. Instead, they wanted to know the answers to the sorts of questions that I've presented as the rightful realm of bioethics: whose interests are served, who is harmed and how can justice be done?

Most people actually expressed their concerns about whether we should do whatever science lets us do in terms of what they identified as past science policy mistakes, such as BSE (bovine spongiform encephalopathy or 'mad-cow' disease), thalidomide or nuclear power. As Kitzinger says, 'Actual history was a more powerful warrant than futuristic fiction.' She concludes that science-fiction fears are what philosophers call a 'straw man': an oversimplified, inaccurate or obviously false argument illegitimately attributed to one's opponent, the more easily to be knocked down.

But fair and rational discussion is all too rare in biotechnology policy-making. When in 2008 the UK was debating a new version of the Human Fertilisation and Embryology Act, the two sides allowed themselves to be drawn into a highly polarized media debate that provided an object lesson in how not to do bioethics. The stumbling block was 'cybrids' or 'human admixed embryos', in which an animal egg is hollowed out to insert human genetic material but not allowed to develop beyond 14 days. This was claimed to be a crucial research technique that would get round the massive need for human eggs in SCNT research. In this case, one opponent – Cardinal Keith O'Brien – did use the 'F' word: Frankenstein. But 'cybrid' supporters presented his remarks as typical of the ignorance behind all attempts to limit whatever science allows us to do, while simultaneously implying – misleadingly – that opposition to this particular form of research would doom all stem cell science.

The great 'cybrid' debate

In one corner, wearing black trunks and a red biretta: Cardinal Keith O'Brien, head of the Roman Catholic Church in Scotland, who alleged in an Easter sermon that the Human Fertilisation and Embryology Bill – now before Parliament – would allow 'government-supported experiments of Frankenstein proportions'. In the other, wearing white trunks and a silver halo: the scientists' champion ... If this sounds like a caricature, that's because it is. But it's no more of a parody than the way in which serious ethical debate over the Bill has descended into a slanging match ...

As a (secular) bioethicist sitting on ethics committees, I find it unhelpful to be called 'Luddite' or 'God-botherer' when I pose ethical questions about new scientific developments. There's a pressing need to address issues that have little to do with whether or not you believe in God.

[T]he commercialised medical environment means that researchers can compete internationally for glittering financial prizes. Vast amounts of money are being poured into private–public collaborations all over the globe. UK

scientists want their share, naturally enough, and in the international stem cell research competition they lead their US competitors, hampered by [former president] Bush's prohibition on federal funding.

... So we need to take a more critical attitude, examining whose interests are at stake – and how plausible a new scientific process really is. Take the case for human admixed embryos, which are supposedly crucial for vital stem cell research. True, last month a Newcastle team announced they'd grown such embryos to the 32-cell stage – but that's a long way from saving lives. As Conservative MP Mark Pritchard has warned: 'There has been no recorded case of a single patient who has been cured of any disease using human embryonic stem cells, a small detail omitted by the large bio-tech corporations that stand to make millions from the Government's proposals'.[14]

In fact, although the revised law did eventually permit 'cybrids', they didn't lead to anything like the progress foretold. The UK Medical Research Council later withdrew funding from most of the projects working in this area, as being scientifically less productive than other forms of stem cell research. That's another reason to ask difficult questions: we can't necessarily be certain that even the most promising techniques will fulfil their proponents' claims for them, and there are institutional or financial temptations for even the most honest researchers to 'hype' their research.

Bioethics has been in the business of asking tough questions since its origins half a century ago in the narrower form of medical ethics, concerned with clinical treatment issues about life-saving treatments. The first technology to be interrogated was kidney dialysis, when

▲ A leader of the Luddites, nineteenth-century English craftsmen who protested against the Industrial Revolution by destroying mechanical looms, dressed as a woman and urging his companions to attack another factory.

in the early 1960s a committee in Seattle, Washington, attempted to set rules for who should receive what was then a scarce resource.[15] But the Seattle 'God Committee', as the group was derisorily labelled, did medical ethics no favours by the methods and outcomes of their deliberations. Using social criteria such as church membership, income and even Scout leadership, the committee discriminated against the less well-off and awarded the life-saving resource of dialysis to men three times more often than to women.

What was wrong with the decisions of the Seattle 'God Committee' was that they reinforced rather than challenged existing injustices – not that members were 'playing God' when they should have avoided making moral judgements. No one can avoid making moral judgements altogether: it was the content and method of the committee's judgements that were faulty. Accusing someone of 'playing God' is itself a moral judgement. As the British medical ethicist Tony Hope has written, we have to decide which actions are right or wrong before we decide what constitutes 'playing God', so it would be circular to say that 'playing God' makes a particular judgement wrong.[16]

In an attempt to rise above such subjective preferences, and as part of the 'social engineering' movement typified by Lyndon Johnson's 'Great Society' (1963–1968),[17] medical ethicists in the USA tried to develop more objective ways of elucidating difficult clinical decisions – particularly end-of-life issues, once new artificial resuscitation techniques could keep patients alive more or less indefinitely. The philosopher Tom Beauchamp and the theologian James Childress[18] developed a very influential set of 'four principles' – autonomy, non-maleficence (doing no harm), beneficence (doing good) and justice. 'Principlism', with autonomy definitely first among not-so-equals, dominated medical ethics in the USA for a very long time and is still highly influential in American and British medical schools today.

Other countries, however, have been much more sceptical about this autonomy-focused model. In France, the first national bioethics committee in Europe, set up in 1983, has instead emphasized solidarity with the vulnerable, justice and human dignity. The French continue to reject 'Anglo-Saxon values', with the most recent revision of their biotechnology legislation (2011) still outlawing commercial surrogacy and egg sale. Values other than autonomy also predominate elsewhere in Europe: for example, the Nordic focus on social provision and community cohesion rather than choice and autonomy.[19] In Tonga and New Zealand, activist movements insisting on indigenous values such as respect for *ngeia*, the principle of life, have rejected outside attempts to commercialize their unique genetic resources.[20]

More generally, there has been a growth in 'global bioethics', with two aims: to foreground neglected issues affecting non-Western countries, such as outsourcing research, and to debate whether there can be any agreement on genuinely universal values. Another powerful alternative voice is provided by feminist critics, who have insisted that autonomy needs to include a 'relational' component to reflect women's lived experience of relationship, and also that bioethicists should be alert – as very few were in the Hwang case – to potential injustices for women.[21]

That's important, because the origins of bioethics also lie in public outrage over injustices. In the 1960s and 1970s, when the discipline was emerging in its earlier form of medical ethics, there were many of those, such as:

❯ The case of the Auckland (New Zealand) Women's Hospital, where women with precancerous cervical abnormalities were studied but not treated for cancer, although effective interventions were available.

❯ The grimly similar case of the Tuskegee, Alabama, 'studies', in which African American men with syphilis were monitored but were deliberately denied clinical care.

❯ The abuses of research ethics in the UK documented in Maurice Pappworth's *Human Guinea Pigs*,[22] including injecting patients with malaria parasites, cancer cells and live polio virus without their knowledge.

Do such abuses still happen? Much less often, in part due to the way in which ethics and law have become a require-ment in medical school teaching – although in 2001 the UK had the Alder Hey Hospital scandal, in which a pathologist retained dead children's tissue without the knowledge or consent of the parents. But bioethics isn't just about reacting to such extreme and atypical cases: it should aim to be proactive, forecasting when and where new biomedical technologies might transgress human rights or discriminate against vulnerable groups.

That might sound obvious to the point of triteness, but some progressive bioethicists are actually very concerned that their discipline has ceased to 'speak truth to power' and has instead become part of the corporate structures of modern commercialized biomedicine. In part this is the cost of the discipline's own success in permeating medical school teaching, consultative soundings, national ethics commissions and media reports. But as the US bioethics professor Carl Elliott writes: 'As bioethicists seek to become trusted advisors, rather than gadflies or watchdogs, it will not be surprising if they slowly come to resemble the people they are trusted to advise. And when that happens, moral compromise will be unnecessary because there will be little left to compromise.'[23]

That's a valid warning against uncritically accepting whatever science throws at us, but we should also be concerned about the converse: a frenzied reaction against any new biotechnology development as 'unnatural'. Whatever is wrong with any new biotechnology, it won't be simply that it's 'unnatural'. The 'unnatural' argument is like the one about 'playing God': it rests on a previous ethical judgement that isn't just self-evident. Preventing women from dying in childbirth and lessening infant mortality are also unnatural, but most of us would deny that either of those endeavours is a bad thing.

> *Science never solves a problem without creating ten more.*
>
> George Bernard Shaw

It sometimes seems as if bioethics, like the bioscience it analyses, 'never solves a problem without creating ten more'. But that doesn't mean we should give up the attempt to apply our best intelligence and most sensitive feelings to these matters of life and death. There's no way we can avoid making moral judgements about the new biotechnologies; nor should we do whatever science lets us do. The title of this chapter isn't just a rhetorical question, because some people really do think that we should. But you'll have spotted some time ago that I don't agree. We can do much better than that. In the remaining chapters, I will try to show how.

2

'Girls! Sell your eggs and enjoy the nightlife of Chennai!'

Egg donors needed to build families. Travel to India. Cash compensation and a three-week trip to India planned around your academic schedule. Ages 20–29 only.

Advertisement in college newspaper at Duke University (Durham, North Carolina)[1]

ALL THAT MATTERS

Along with the small ads for local pizza parlours, advertisements like this are increasingly common on American campuses. They're the local outcroppings of the 'reproductive tourism' iceberg: the global trade in eggs, sperm and 'surrogate' motherhood that has developed in the 30 or so years since the birth of Louise Brown, the first baby conceived through in vitro fertilization (IVF). Those who think that we should let science do whatever it can do would agree with political libertarians that, although the advertisement looks slightly risible, there's nothing wrong with paying young fertile women to 'donate' eggs, either at home or abroad. Others might think the practice should be regulated to prevent the worst abuses – unlicensed clinics, say, or excessive hormonal regimes to stimulate dangerously large quantities of eggs – while more strenuous critics would regard 'reproductive tourism' as needing to be banned altogether, because it represents the purchase of fertility from poor women.[2]

Examples of all three approaches can be found both within the USA and across the world,[3] but those who would abolish or even merely regulate the international fertility trade often feel that they have an uphill struggle against the interests of a highly profitable global business. Those interests are hardly challenged by simplistic but influential reactions like the following:

▶ 'That doctors would be so paternalistic as to deny women the option of using a surrogate if the surrogate were willing to do so is simply outrageous.' (from a US bioethicist).[4]
▶ 'Everyone – including ugly people – would like to bring good-looking children into the world, and we can't be

Girls! Sell your eggs and enjoy the nightlife

22

selfish with our attractive gene pool.'[5] (from the founder of the website 'Beautiful people', which sells eggs and sperm from beautiful women and handsome men).

▶ 'If I can give my kidney to someone who needs it, why not give a baby to someone who can't have one?'[6] (from a paid American 'surrogate' mother).

These statements have one thing in common: they reduce a complex ethical debate to a matter of individual choice. The first statement tries to use paternalism as a knock-down argument, permitting no further discussion. That style of argument is always suspicious: in my opinion, bioethics should be about opening up minds and questions, not shutting them down.

The second statement self-evidently treats the decision to bring a child into the world as a consumer choice. You wouldn't buy an ugly old banger if you could have a sleek and stylish new sports car, so why settle for an ugly child – even if you have the appalling misfortune to be less than stunning yourself? At least, that's how this thinking runs.

The third statement looks less self-absorbed on the surface – even altruistic. However, apart from the fact that this transaction is not actually gift but sale, and that even in the USA you're not allowed to sell a kidney, this 'surrogate' is leaving one thing out: a kidney isn't a person. She is making a decision that will affect not only her own future, but that of the baby.

That may sound obvious, but it is often ignored, perhaps because it threatens to undermine the argument from paternalism (as in the first quotation). Once the interests of a person other than the commissioning couple and the 'surrogate' are included in the moral equation, it's not enough simply to call any interference paternalistic. John Stuart Mill, the nineteenth-century philosopher widely venerated by libertarians, himself wrote, 'The only purpose for which power can be rightfully exercised over any member of a civilised community, against his will, is to prevent harm to others.'[7] So even a libertarian needs to acknowledge the possibility that the baby is harmed by gamete donation or 'surrogacy': we can't simply rule out that question as paternalistic.

In 'surrogacy', the resultant child can be harmed by being left parentless if the commissioning couple reject the baby because it is the 'wrong' sex or disabled, or if the baby buyers come from a jurisdiction that doesn't recognize 'surrogacy' as legal, meaning that the child is denied the right to enter the country and is left stateless. In one case, an Italian man and his Portuguese wife arranged for a British 'surrogate' to be inseminated in a Greek laboratory, using sperm from an anonymous American donor, who had contributed to a Danish sperm bank, and taking eggs from another British woman. The commissioning couple then refused to accept the twin girl babies because they were the 'wrong' sex, and so the children were sent for adoption.

Paid 'surrogacy' hasn't been around for long enough for us to have a pool of adults born through the process, but we do have some systematic surveys and testimony from young people born through egg or sperm donation. In its 2004–2006 review of whether gamete donors should be paid wages or increased expenses, the UK Human Fertilisation and Embryology Authority found that 'all the

Girls! Sell your eggs and enjoy the nightlife

24

▲ The nineteenth-century philosopher John Stuart Mill thought liberty paramount unless others may be harmed – which could rule out baby selling or 'surrogacy'.

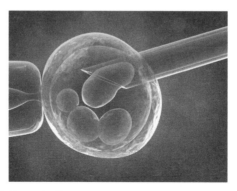

▲ Artificial insemination

donor-conceived people who replied thought that even to pay expenses would risk obscuring the donor's motivation.'[8]

An American survey of donor-conceived adults, 'My Daddy's Name is Donor',[9] reported that about half the respondents found the circumstances of their conception disturbing. Its co-author has since written, 'In our opinion, an elective procedure used to treat one person's medical or social issue, which has the most direct effects on another, entirely different person who is, at the time, unable to speak for him- or herself or consent to treatment, should be held to the rigorous ethical test of asking, "Is anyone harmed at all?" In the case of donor conception, our study and ample other work are showing that indeed the resulting persons can be harmed by this practice.'[10]

▲ A fertile woman forced to be a surrogate mother ('handmaid') from the film of Margaret Atwood's 'speculative fiction', *The Handmaid's Tale*.

Girls! Sell your eggs and enjoy the nightlife

26

As a British donor-conceived adult wrote in response to an article by an advocate of paying for gametes:

> *I am a person born from sperm donation. I am disappointed although not altogether surprised that the author did not consider a crucial question. How will people born from these arrangements feel once they are old enough to understand that their biological father or mother only did it for the money? I have friends conceived from donated gametes and we all agree that it is dehumanising to have your life created from a commercial relationship. If fertility treatment is big business and operates under business principles of prices regulated by supply and demand, I guess that makes us the 'products'.[11]*

Making the formerly infertile fertile is one of the genuine advances in modern biomedicine, but it doesn't benefit everyone equally. Issues about justice and exploitation will be at the forefront in this chapter, and, as always with issues about justice, the question will be whether equals are being treated unequally. The first step is counting everyone affected – babies as well as gamete providers, 'surrogates' and commissioning couples. But who counts as equal, and what counts as treating them equally? Are buyers and sellers in different bargaining positions? What counts as rightful reproductive autonomy, and what practices are impermissibly exploitative of someone else's weak position?

A phrasebook for 'reproductive tourists'

The terms I need to use in this chapter are contentious, but that isn't just an academic gripe: they often add a spurious legitimacy to practices that we ought to be interrogating as ethically questionable. Here is a list of some terms used commonly in reproductive bioethics, along with reasons for dissatisfaction with them and possible alternatives.

Egg donor When, in return for her eggs, a woman receives cash or a benefit in kind such as discounted fertility treatment ('egg sharing'), that sounds like a sale to me. The language of altruism ('donor') camouflages the transaction. Since the influential comparison of the UK and US blood donation systems by Richard Titmuss in his 1970 book *The Gift Relationship*,[12] the language of gift has been press-ganged into service in some rather dubious ways. In the extreme, one commercial agency refers to the women who supply eggs as 'donor angels'.[13] 'Egg seller' is my preferred term, unless neither payment nor 'egg sharing' (itself another euphemism) is involved. 'Egg supplier' is a neutral possibility that would also encompass genuinely altruistic donation.

Surrogate motherhood The common law of the English-speaking countries holds that the woman who gives birth is the mother. There's nothing 'surrogate' about it, although, more recently, individual cases have begun to undermine this principle.[14] Nor does the law depend on whether the implanted embryo is created from the birth mother's own eggs or someone else's, although some writers use the terms 'partial surrogacy' and 'full surrogacy' to distinguish between the two situations. The term 'gestational carrier' is increasingly de rigueur: in her account of commissioning 'twiblings', fraternal twin embryos implanted in two separate birth mothers (Melissa and Fie), Melanie Thernstrom complains, 'I was irked by all the people – especially health care professionals – who were unable to master the term "gestational carrier" and referred to Melissa and Fie as "birth mothers" or "biological mothers" ... Everything about third-party reproduction can suggest you are not the "real mother".'[15] The only surprising thing about it is that Thernstrom is surprised: that has been the legal position for centuries. 'Biological mother' is itself

Girls! Sell your eggs and enjoy the nightlife

28

ambiguous, because it could mean either the genetic mother or the birth mother. 'Contract pregnancy' and 'contract motherhood' are used by some analysts, but commercial surrogacy contracts are not legally binding in many US states and most European countries, meaning that any arrangement made is not really a contract. 'Pregnancy outsourcing' is politically savvy, but what's being paid for is the baby rather than the pregnancy. That other monetary description, 'renting a womb', is not only flip but grossly inaccurate, omitting as it does the risks and pain of childbirth. The best term is the one few will tackle: 'baby-selling'.

Reproductive tourism Although advertisements like the one in the Duke campus newspaper deliberately play up the tourism angle, some commentators think that this term is too flippant to describe the trials endured by infertile couples. The British fertility specialist Francoise Shenfield prefers the term 'cross-border reproductive care'; she also dislikes the term 'global fertility industry'. However, she ruefully acknowledges an inconvenient truth when she writes, 'If we fertility specialists are branded an industry by some, it must mean the image we sometimes give is a money-making machine, not a professional medical group working to standards.'[16] Shenfield would presumably be even more disturbed by the terms 'baby markets'[17] or 'the baby business',[18] both used in the titles of recent academic volumes by well-regarded experts, some of whom actually view the terms as favourable. The US law professor Martha Ertman, for example, happily uses the term 'baby markets' in her article 'The upside of baby markets', writing: 'Most people object to markets in babies. I disagree ... because market mechanisms provide unique opportunities for law and culture to recognize that people form families in different ways.'[19]

Are those young women who are tempted to 'sell their eggs and enjoy the nightlife of Chennai' liable to be exploited and put at medical risk? Or is this a win–win situation in which the women can pay off their college expenses, while childless Western couples get the 'egg donors needed to build families'? (That appeal to altruism comes first in the advertisement, interestingly enough, before the more mundane lures of cash and vacation.) Some of those couples will also employ Indian 'surrogate mothers', a growth industry that the Indian government is seeking to put on a solid legal footing so as to maximize foreign currency earnings. Don't those mothers benefit from payments far in excess of what they or their husbands could ever earn in a lifetime?

> I deliver babies from Indian women for British couples at the rate of more than 13 a month. For these surrogate mothers, the amount of money is life-changing. It helps them set up a home or get their daughters married. There is absolutely no exploitation of these women. It is really big money. It is a jackpot. They go through a little bit of emotional trauma, but then they go back home and realise they have done it for a good cause.

Dr Anita Soni, a fertility specialist at a leading hospital in Mumbai[20]

That 'little bit of emotional trauma', of course, is surrendering the baby to whom you've given birth. Nevertheless, isn't it the mother's choice? Many of the women involved robustly deny that any exploitation is taking place in the surrogacy clinics:

> This is not exploitation. Crushing glass for fifteen hours a day [her usual job] is exploitation. The baby's parents have given me a chance to make good marriages for my daughters. That's a big weight off my mind.'

An Indian woman having her sixth child, her first for payment[21]

Crushing glass for 15 hours a day *is* exploitation, but that doesn't itself make 'surrogacy' exploitation-free. (The British global ethicist Heather Widdows argues that it's extremely exploitative, even coercive, to take advantage of someone's lack of other choices for keeping body and soul together.[22]) Neither does the idea that it's better than prostitution, although one Indian infertility doctor uses exactly that argument:

> *To convince the women I often explain to them that it's like renting a house for a year. We want to rent your womb for a year, and Doctor Madam will get you money in return. I tell them surrogacy is not immoral. It is much better than a woman going from one man's bed to the next to make money. Prostitution will not pay her much and can also lead to diseases.*[23]

The suspicion that the birth mothers' autonomy isn't really the name of the game is reinforced by reports of the way in which some of these clinics physically confine the women. A widely quoted article on a clinic in Gujarat claimed that the women were effectively imprisoned for nine months, only permitted to leave the premises for hospital check-ups and confined to an upper-floor room lined with eight to ten single iron beds, with barely enough room to walk between them.[24] Yet the clinic's director insisted that the 'surrogates' enjoyed staying at the clinic: 'For the women it's like a paid holiday.'[25] (Funny how the vacation theme keeps cropping up.)

The autonomy of the Indian birth mother is also severely constrained by the requirement in the new legislation that she must hand over the baby, by law. When a 'surrogacy' contract mandates that the baby must be delivered to the 'commissioning couple', the practice looks parlously close to baby-selling – effectively a form of slavery. It is quite surprising how readily recent bioethical debate has ignored that minor quibble (although it was part of the UK Warnock report on artificial reproductive technology in 1985).

Even Debra Satz, a perceptive and robust critic of commercial 'surrogacy', dismisses the argument that contract motherhood is equivalent to baby-selling. She remarks that '[T]his argument is flawed. Pregnancy contracts do not enable fathers (or prospective mothers, women who are infertile or otherwise unable to conceive) to acquire full ownership rights over children. Even where there has been a financial payment for conceiving a child, the child cannot be viewed as a mere commodity. The father (or prospective mother) cannot, for example, simply destroy or abandon the child.'[26]

Property rights are viewed in law not as a single all-or-nothing criterion, but as a 'bundle' of entitlements and duties about an object, X.[27] We can have some, all or none of the sticks in the bundle, which may include:

- ❱ the right to physical possession of X
- ❱ the right to its management, for example to determine the ways in which others can use it, or even to destroy it
- ❱ the right to security against its being taken by others
- ❱ the right to transmit X to others by sale, gift or bequest.

But actually it's Satz's argument that's flawed, because she's thinking of property as unitary. While I have a property in my vote, for example, I don't have the right to sell it. Similarly, while the commissioning couple don't have the

Girls! Sell your eggs and enjoy the nightlife

32

▲ An Indian clinic for surrogacy.

right to destroy the 'object' X – the baby – a surrogacy contract does specify other rights in the property bundle: physical possession, management, security against the birth mother claiming custody, and presumably the right to provide in their wills for the child's ongoing care, which is a sort of right to transmit to others. (I want to make it clear that I don't think that the child is an object – but pregnancy contracts treat it as one.) Hiring a contract mother is sometimes presented as a 'service', like prostitution, but that's inaccurate: the contract is for the product of the pregnancy, the baby – not for any benefit or pleasure that the buyer reaps from the pregnancy process itself.

So it's certainly arguable that surrogacy contracts do treat the child as an object of property and therefore equate to baby-selling – particularly when, as in the case of the Indian surrogacy bill, the government puts its weight behind them, making them legally enforceable like any other property claim. In fact, the proposed law's compulsion on the birth mother to hand over the baby actually exceeds the 'specific performance' claims of other contract rights.

As Satz correctly points out, if I pay you to paint my house but you renege on the deal, the law can't physically force you to scale a ladder and get busy with your brushes; all it can compel you to do is return my money, possibly with damages. Not so with surrogate mothers in India: under the proposed legislation, they can be forced to complete the 'job' by handing over the baby – not just returning the money and keeping the child.

These infringements of the birth mother's autonomy are a powerful counterweight to the argument that, if a woman protests that she is not being exploited by 'surrogacy', we must take her at her word or risk being paternalistic. The interests of the future child, already explored earlier in this chapter, also constitute an important check on the argument from paternalism or its variant, neo-colonialism (which regards Western critiques of Indian biomedicine as unwelcome warmed-over imperialism). But I'd like to go further than that. In this final part of the chapter I want to ask whether we can regard commercialized gamete provision and contract pregnancy as wrong, even apart from the question of the child's interests, and even when the women involved have given their consent.

There are two possible reasons for thinking so. The first is the way in which these practices commodify the body, turning it (or, more accurately, parts of it) into an object

Girls! Sell your eggs and enjoy the nightlife

34

of trade. In our traditional common law, the body wasn't regarded as property: tissue separated from the body was regarded as *res nullius*, no one's thing, and in fact not really a thing at all. That position chimes with the clear statement by the eighteenth-century philosopher Immanuel Kant that something may be either a person or a thing, but not both. Since we as persons inhabit our bodies, they are not things like any other consumer good. In the words of a more modern philosopher, Maurice Merleau-Ponty:

> If the body is a thing among things, it is so in a stronger and deeper sense than they.[28]

Although Kant is usually cited as the key source for the importance of autonomy, he very specifically denied that we could autonomously choose to treat our bodies merely as means to an end, such as raising money. That was self-contradictory, he thought: an agent can't rationally set out to treat himself as an object. But Kant himself had problems with excised parts of the body. He acknowledged that it might be permissible, although not virtuous, to sell one's hair, for example.

So why not sell eggs? And if we are allowed to sell the labour of our bodies to the highest bidder, how is contract pregnancy any different? Those who favour these practices argue that the body and its labour are already commodified in modern capitalist society. Some might also assert that both practices actually recognize that women have a property in their reproductive labour, and that they are liberated by having the freedom to sell that labour on the free market. Disagreements have arisen between those feminists who believe that the obligation to surrender the baby must be honoured as the downside of recognizing women's property in their own reproductive labour and those who think that pregnancy contracts are inherently exploitative.[29]

I have argued elsewhere that women's reproductive labour is rarely recognized or valued.[30] But, ultimately, I think that the former viewpoint is naive, and that leads into the second strike against commercial surrogacy and egg donation: women do not and cannot consent freely in these decisions, making them open to coercion or exploitation. Often they lack sufficient information to give an informed consent: many of the Indian women surveyed did not understand that IVF would be necessary, with its attendant risks for both donors and recipients of eggs. Being told that Doctor Madam wants to rent your womb for a year isn't anything like enough for genuinely informed consent.

As far as egg sales are concerned, although we live in an era that stresses the need for the medical evidence base, there are very few long-term studies of what risks women undergo in egg retrieval, and very few attempts to set any up.[31] One study has shown that some private US clinics provide grossly inadequate and even false information on risks to egg sellers.[32] We do know that the procedure can take up to 60 hours, that the drug used to induce extra eggs has caused fatalities and that attempts by the US professional body to regulate amounts paid are widely flouted.[33] So it's not just a holiday in Chennai.

> [H]ow many people know that obtaining eggs for in vitro fertilization requires women (whether

undergoing fertility treatment themselves or providing eggs for others) to take a heavy load of hormones; that some of these drugs have not been approved by the FDA [US Food and Drug Administration] for this use; that there have been thousands of reported adverse reactions to the most commonly used of these drugs, including hundreds of hospitalisations; and that no one is systematically collecting data on the long-term effects on women and their offspring? ... In essence, by allowing the fertility industry to experiment with new techniques and protocols with little oversight, and by uncritically embracing these new technologies, we have put women and children at risk and crossed numerous moral and ethical lines. Moreover, these lines have been crossed with little public acknowledgement. We believe that it is time for our community to undertake a pro-active, in-depth, critical analysis of the safety concerns and ethical dilemmas posed by new reproductive and genetic technologies.[34]

The philosopher Zahra Meghani argues that we have to understand individual choices in the context of global neo-liberalism and its associated policies: privatization, deregulation and commodification.[35] Rather than an apolitical, one-size-fits-all argument such as 'choice', we also have to understand local realities. The feminist group Sama puts in a similar plea for understanding the Indian 'surrogacy' situation in its social context. On that basis Sama opposes the draft surrogacy bill – reiterating in its statement a theme from Chapter 1 of this book: that we shouldn't passively accept whatever science makes possible.

> [E]verything that is medically possible should not necessarily be legally permissible. Law is an instrument of social engineering and must be developed with consideration for all sections of society, especially those that are more vulnerable and marginalised, to prevent any kind of exploitation.[36]

So much for the charge that those who criticize the Indian 'surrogacy' situation are neo-colonial paternalists: Sama is itself an Indian group. Rather, the genuine new imperialists are those exporting the individualistic 'choice' ideology from West to East. Neo-liberal policies – rolling back state provision – have often taken public healthcare provision away from many deprived lower-caste rural Indian women (the bulk of 'surrogates') and are pushing even comparatively affluent US and UK students more and more into debt. To suggest that there is a possibility of exploitation in their 'choice' to redress their economic situation as best they can – by egg- or baby-selling – is not paternalistic, just realistic.

The very notion of 'choice', so foundational to bioethics, has itself been quietly transformed by neo-liberal ideologies, as the US bioethicist Lisa Ikemoto argues. While choice was once a genuinely liberating argument, conferring greater power on patients or enabling women to obtain abortions, it has now become a mere prop for 'free market' ideology in reproductive bioethics, she thinks:

> The linkage between choice, autonomy and equality was reworked into an understanding of reproductive choice as an aspect of free-market individualism ... Bodily integrity, decisional autonomy and equality were replaced with free-market individualism and ownership.[37]

In Chapter 1, we saw that the argument about an altruistic duty to contribute to research ignored the political reality of privatized research delegated to profit-making companies. Here in Chapter 2, we've seen that another policy from the same stable, commitment to free global trade unhindered by regulation, has resulted in an explosion of 'reproductive tourism', most pronounced in the most neo-liberal and most technologically advanced of developing countries, such as India.[38]

Some writers acknowledge that truth but still think the trade just needs to be regulated.[39] The legal scholar Margaret Radin has written, 'We can both know the price of something and know that it is priceless.'[40] Another academic, Casey Humbyrd, extends this logic to argue that the only ethical problem is not whether the children have a price on their heads but whether the mothers are being treated fairly. But this seems very much of a sentimental camouflage to me – if the mothers are paid for the babies, the babies are not priceless.

Just as we have 'fair trade' chocolate or bananas, Humbyrd has argued, so we could have 'fair trade surrogacy'. But babies aren't bananas: they're people. You can't have 'fair trade' people.

3

Designer babies, transhumans and lesser mortals

Why be you, when you can be new?

From the film Robots

ALL THAT MATTERS

If you could be new in any one way – if you could 'upgrade' your natural traits and powers with one single extra ability – what would you choose? Survey groups of UK school pupils were asked that question after viewing films, such as *Robots*, about biotechnological 'enhancement'.[1]

"If it upsets you, don't read it."

▲ Would humans be more rational if they became more like robots?

Enhancement

'Enhancement' has been defined as 'a deliberate intervention, applying biomedical science, which aims to improve an existing capacity that most or all normal human beings typically have or to create a new capacity, by acting directly on the body or brain.'[2] These speculative technologies include neurocognitive stimulation techniques, drugs and genetic manipulation – either to your own body, or, more controversially, germline genetic modification affecting your descendants as well. Where therapy ends and 'enhancement' begins is a controversial issue, with the British sociologist Nikolas Rose warning that 'The old lines between correction, treatment and enhancement can no longer be sustained'.[3]

Most of the younger children chose flying, as I would have done myself at their age. (I might think better of it now that I know, from *Harry Potter and the Deathly Hallows*, that the only other human who can fly unassisted is Lord Voldemort.) But older and possibly wiser heads among the pupils rejected the question altogether:

> *In terms of upgrades that enhance beyond human capacity, I would not require any upgrades. The search for perfection is created by a desire to be 'better'. This can never be fulfilled, in a sense. What is important is to be content with one's current situation.*
>
> *17-year-old boy*

Even among the younger children there were some serious sceptics, such as the 11-year-old boy who wrote:

> *I don't want any upgrades because it [is] imorle [immoral] to change who you are. You are born and your [sic] special, if you upgrade humans everyone would want the same thing, it's stupid.*

Except for a few grammatical niceties, there's not a great deal of difference between this boy's thinking and that of one of the most prominent opponents of enhancement – the German philosopher Jürgen Habermas, who has said: 'Interventions aiming at enhancement ... violate the fundamental equal status of persons as autonomous beings ... barring [the 'enhanced' person] from being the undivided author of his own life.'[4]

Both Habermas and the 11-year-old boy stress uniqueness and personal autonomy. If biotechnological enhancement were possible – and it's important to note from the start that its likelihood, like the report of Mark Twain's death, has been greatly exaggerated – it might undermine the 'specialness' of each person. But the 11-year-old actually goes further than Habermas in raising the related risk of eugenics: 'If you upgrade humans, everyone would want the same thing.'

There's already some evidence of this phenomenon in egg sales, in which conventionally 'desirable' traits such as tall height, high intelligence and blond hair dominate most of the advertisements in US egg markets. But what differentiates this phenomenon, and enhancement, from eugenics as practised by the Nazi regime is that it's not a governmental policy. No deliberate social engineering is involved: convergence towards a uniform pattern of physical appearance or genetic characteristics would occur without co-ordination, through the separate choices of many individuals. So although eugenics is the objection to enhancement that occurs first to most people, it's actually not a fair or accurate charge – even though one strong advocate of enhancement for anyone who can afford it has defiantly titled his book *Liberal Eugenics*.[5]

Likewise, it's not really correct to talk about 'designer babies', the possibility (exaggerated by the media) of creating children to order. It's true that 'one-stop baby shops' have sprung up in California, Texas and elsewhere, allowing prospective purchasers – I use the word advisedly – to select particular characteristics from the egg and sperm sellers.[6] If that phenomenon were to spread, and if there were convergence on a preference for blond, tall, musical, intelligent and athletic egg or sperm sellers, then we might be witnessing something like a eugenic phenomenon created solely by individual choice. But there is also evidence of particular preferences that break the eugenic mould, particularly – and ironically, given the Nazi atrocities – for Jewish women's eggs.[7]

There's another crucial difference. Pre-implantation genetic diagnosis (PGD), the IVF-related technology used to create so-called 'designer babies', originated in a very different motive from eugenics: not engineering socially desirable traits, but avoiding medically harmful or even fatal conditions (see box). There are two key differences here: social versus medical judgements, and positively trying to create a desired set of traits versus negatively seeking to avoid harmful ones.

Because of those vital distinctions, and because PGD is not undertaken lightly, I think the label 'designer babies' is unhelpful, snide and flippant. It underestimates and undervalues what women go through: the hazards of ovarian stimulation, egg retrieval and implantation of a selected embryo after fertilization and elimination of 'defective' embryos. PGD typically means discarding embryos with the harmful condition, raising fatal ethical objections for those who believe that life begins at conception. (This is why the procedure is banned in some countries.) Even for those who, like me, do not believe that life starts at fertilization, there are still serious moral questions involved.

Techniques for embryo testing

PGD is a technique in which couples at known risk of carrying a serious genetic condition undergo IVF to identify embryos without the condition. For example, PGD can be used for inherited conditions such as cystic fibrosis or Tay–Sachs disease (a genetic illness, predominantly affecting Ashkenazi Jews, in which most children die before their second birthday). These are both recessive conditions, in which the mother and father may appear healthy because they only have one allele (genetic variant) for the illness. But if their offspring inherit one such allele from each parent, they will have the disease. If two parents are both carriers of the harmful variant of the gene, but do not display the condition themselves, on statistical average one of every four embryos would manifest the disease, two would produce 'carriers' like the parents, and one would be completely free of the condition, neither a carrier nor a sufferer. The hope would be to produce at least one embryo in this last category and replace it in the uterus.

In the UK PGD is allowed, but it is not permitted to implant a 'defective' embryo. However, there has been controversy about whether genetically linked deafness is a 'defective' condition, with some members of the Deaf community actively preferring to raise a Deaf child as part of their culture. They view deafness as an identity rather than a medical abnormality.[8]

For affected couples, the alternatives to PGD (besides remaining childless, adopting a genetically unrelated child or accepting the risk and letting a naturally conceived pregnancy take its course) are either aborting a normally conceived fetus later found to carry the harmful variant or non-invasive fetal testing, a recently developed technique requiring only a maternal blood test. This could be done at a much earlier stage of the pregnancy. In December 2010 it was announced that an entire fetal genome had been screened, from the mother's blood sample, for beta-thalassaemia (a serious genetic condition affecting many people of Mediterranean origin).[9]

That's one thing, but what about lesser abnormalities? Given that most of us carry at least six or seven adverse alleles, a full fetal

DNA profile is almost bound to reveal some serious problems for many more couples than the comparatively limited number for whom PGD was originally designed. As the US activist and analyst Marcy Darnovsky asks, 'How could pregnant women and their partners possibly interpret the results of tests that claim to predict dozens or hundreds of a future child's traits?'[10]

All this assumes that the couple already know that one or both of them would be at risk of passing on a serious genetic condition to their offspring. That raises the issue of whether whole-population genomic screening is desirable. Britain's Human Genetics Commission concluded in April 2011 that there was no ethical reason why it *shouldn't* be inaugurated, but that's a long way from concluding positively that it should be.[11] However, there are precedents: in Cyprus a population-wide screening programme requires couples to present a certificate of testing for beta-thalassaemia before being allowed to marry. Those who test positive as carriers are still permitted to wed, but many opt for prenatal diagnosis or abortion, with the unexpected approval of the Greek Orthodox Church. Political and economic arguments have won the day: it was predicted that, without such measures, the country's entire health budget would be spent on beta-thalassaemia within 40 years.

Somewhere between the lofty plateau of obviously 'medical' motives for selecting one embryo over another and the murky bog of clearly 'social' ones is located the terrain of sex selection. On one border of that contested territory lie sex-linked conditions such as haemophilia, in which normally only boys manifest the full condition. (Also a recessive disorder, haemophilia is linked to a mutation on the X chromosome: because females have two X chromosomes but males have one X and one Y chromosome, boys who inherit one adverse allele are unprotected by the second 'healthy' X chromosome found in girls.) Here sex selection may seem much more justified than at the territory's other flank, sex selection for 'family balancing' or for cultural reasons, such as a historical preference for males. But strong advocates of reproductive autonomy argue that both medical and social choices in favour of sex selection should be permitted freely.

As always with the reproductive autonomy argument, however, it's impossible to ignore or avoid the social consequences of individual choices. In India the increasing availability of sex determination techniques has led to family pressure for women to abort female fetuses: part of the Asian-wide 'epidemic of gender selection' in which, it is claimed, even ten years ago there was already a shortfall of '50 million missing women'.[12] Legislation prohibiting the practice of sex-selective abortion has largely gone unenforced, and a lucrative trade in sex determination and abortion has sprung up. Even in the developed world, things have got out of control: in Canada the demand for sex-specific abortion, more often for 'family balancing', is now so high that a bioethicist and a doctor – both supporters of abortion rights – have recently recommended that doctors delay giving information about the sex of the fetus.[13]

Feminists have faced a dilemma over this issue: if women should have a free choice about abortion, does that extend to the free choice to abort girls for social reasons? But is it really a free choice, or made under threat? On the other hand, is any actual woman harmed by the shortfall? The '50 million missing women' were never born, so how can they have been harmed?

Once again, we're back to the question of whether we should let biotechnology do whatever it can do. Where can we draw a plausible line? If we weigh up benefits and harms, it seems reasonable to think that there could be a negative duty to prevent the birth of a child certain or even highly likely to be afflicted with a fatal medical condition. Sex selection for medical purposes would be allowed under that rubric. Social sex selection, however, requires a healthy woman with no infertility problems to undergo IVF, including the pain, expense and risk of egg collection and embryo transfer, only for the purpose of choosing the baby's sex.[14]

Even if the woman is genuinely willing to undergo these risks – if there is really no family or societal pressure – it is a separate matter whether doctors should expose her to them. Classically, the first duty of a doctor is 'do no harm', and that duty outweighs any non-medical benefits. In a patriarchal society, however, doctors may feel that agreeing to a woman's request to abort a female fetus is protecting her from harm at the hands of her relatives. Those societies are not confined to the Third World: I have heard British obstetricians make that argument about their own patients.

In that sort of society, it might even be said that preferring male fetuses over female gives the future child the chance of a better life. In a strange and repellent way, it would fit the requirement, advocated by the Australian bioethicist Julian Savulescu, to produce the best children we can[15] – although he himself would probably not accept that example. Savulescu, a leading proponent of enhancement as applied to adults, also (and quite consistently) favours using genetic tests to produce those children who, of all possible children whom we could create, will have the best opportunity of the best life. In fact, he goes so far as to say that it is unethical for parents not to do so.

Savulescu does not argue that any particular child is harmed by being born with a less-than-optimal complement of genetic characteristics. That would be illogical: for this particular child, there is no possibility of having been born with a different set of characteristics. That would have been a different child. But he still feels that parents have a duty, which he calls 'procreative beneficence', to bring the best possible children into the world. As a philosophical utilitarian, he is concerned with maximizing the total amount of welfare in the world, and this, he claims, requires us not to knowingly produce children whose existence would lessen the welfare of humanity.

Utilitarianism

Utilitarianism is a sub-branch of the wider ethical school known as consequentialism, which judges the moral value of actions by their beneficial or harmful consequences. General welfare is the consequence that most modern utilitarian consequentialists seek to maximize; others – including utilitarianism's founder, the eighteenth-century philosopher Jeremy Bentham – have emphasized pleasure or happiness.

But Savulescu is actually making a crippling demand, typical of what has been called the oppressive and obsessive regime of utilitarianism. According to one such critic, the American academic lawyer Charles Fried:

> For utilitarians there is always only one right thing to do, and that is to promote in all possible ways

at every moment the greatest happiness of the greatest number. To stop even for a moment or to rest content with a second best is a failure of duty.[16]

Failing to undergo every possible genetic test would presumably count as resting content with a 'second-best' child. The medical reality of Savulescu's position is that it would impose a frightful burden on women: for every pregnancy, either to undergo IVF and PGD, so that the most advantaged embryos can be selected, or to abort fetuses later found to be suboptimal. There is the additional issue of the risks of genetic tests themselves – less serious in the case of non-invasive fetal testing, but significant in PGD (because of the concomitant risks of the IVF that's required).

There are also risks in amniocentesis, sampling of the amniotic fluid through insertion of a needle through the woman's abdomen in the second trimester of pregnancy, to determine fetal abnormality. In another article, 'Parental choice? Letter from a doctor as a dad', Savulescu in fact describes how he and his wife, as self-avowedly 'very risk-averse' and clearly conscientious expectant parents, decided that they would not opt for amniocentesis, once they learned that it was four times more likely to result in spontaneous abortion than in detection of an abnormality.[17]

Women already report feeling under pressure to undergo strenuous and sometimes risky tests. One who refused was Joanna Richards, who has since written:

After Saskia was born [with Down's syndrome, further complicated by brain damage after heart surgery], several people asked me: 'But didn't you have the test?' Somehow, the implicit judgement

> within the question did not fully strike me at the
> time. Looking back, however, I am acutely aware
> of just how value-laden a question it is, translating
> only too easily into: 'You should have had the tests
> and terminated the pregnancy. I judge that you were
> wrong to have given birth to this baby.'[18]

At the same time that Savulescu's 'best child we can' position demands too much of the mother, it also provides too scanty an insight into what constitutes a good life for the child. As my colleague Michael Parker has written in response to the notion of 'procreative beneficence':

> Complex concepts, like those of the good life, the
> best life, and human flourishing, are not reducible
> to simple elements or constituent parts which
> might be identified through the testing of embryos.[19]

Parker's point is trenchant and obvious; yet a similar truth is often ignored by proponents of enhancement. Even if we had a reliable method for identifying the complex genetic components of intelligence and of 'enhancing' them, for example, that would not necessarily produce wiser or more flourishing human beings. (Forty years in academia have convinced me that, while intelligence and wisdom are not necessarily opposites, they are certainly not synonyms.) More profoundly, the French philosopher Michèle le Doeuff questions whether intelligence is an endowment of nature or an attribute of our own making through our life experiences. As she (intelligently) puts it, 'We don't just receive intelligence, we create it for ourselves.'[20]

The US philosopher Allen Buchanan – a careful moderate who describes himself not as straightforwardly 'pro-enhancement' but only as 'anti-anti-enhancement'

– has dismissed concerns that we are not intelligent enough to be sure that any 'improvements' we make will be genuine advances and not irreversible disasters. This is certainly a more sophisticated position than simply attempting to define the difficulty away:

If it wasn't good for you, it wouldn't be enhancement.[21]

Buchanan acknowledges our poor track record in making decisions that actually benefit ourselves, but considers that to be an argument for rather than against enhancement. In terms of our manifold cognitive biases and judgemental errors, things can only get better, he thinks.

The difficulty here is that it's our muddle-headed present selves who are in charge of designing the 'enhancements'. You remember them: the ones whose thinking is so foggy that they need radical help, possibly extending irrevocably even to the genomes of their descendants. They're also the ones who are so prone to look for a technological fix when things go wrong – and enhancement is nothing if not a technological fix. Buchanan appears prey to that reasoning himself, as when he suggests that one genuine enhancement would be the ability to tolerate more extreme fluctuations of climate and temperature caused by global warming. The obvious retort is that it would be better to make a last-ditch stand against global warming, rather than trust that the technological hubris that got us into this mess can get us out again.

There is a similar but perhaps even more troubling question about our inability to predict not just the cognitive make-up of the 'transhumans' or 'posthumans' created by such massive interventions, but also their moral sensibility. In Aldous Huxley's *Brave New World*, moral insight has been enhanced, as straightforwardly as laser eye surgery can now enhance physical eyesight. Rather than wearing their moral beliefs on their sleeves, denizens of the future carry them around in the form of 'soma' tablets. As the character Mustapha Mond explains, 'Anybody can be virtuous now. You can carry at least half your morality round in a bottle. Christianity without tears – that's what soma is.'[22]

But can moral insight ever be enhanced in this or some other fashion? What if the newly 'enhanced' have a hostile

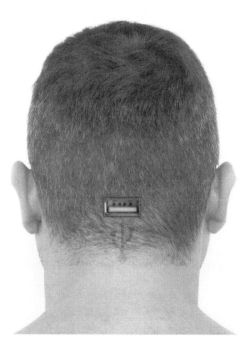

▲ Would this be an enhancement?

set of values to our own? Critics of enhancement stress the likelihood that the 'enhanced' would constitute a powerful new social elite, and the risk that they would have very little regard for the unenhanced underclass. Buchanan quite rightly notes that we can't know whether this would happen: 'Even if biotechnology eventually yields enhancements that are so radical as to call for a new, higher moral status category for the enhanced, the moral status of the unenhanced would not thereby be diminished.'[23]

That's perfectly plausible in terms of our existing concept of human rights as universal, but we can't predict what judgements about moral status the 'posthumans' might make. Given that they've been engineered to be 'superior', they might not be all that charitable. The American bioethicist George Annas warns that '"improved" posthumans would inevitably come to view the "naturals" as inferior, as a subspecies of humans suitable for exploitation, slavery, and even extermination.'[24] That 'inevitably' can't be proven in advance, but do we really want to take the risk? Even Buchanan admits that 'The history of racism and of our treatment of "lower animals" and mentally disabled human beings indicates that [this concern] is not to be taken lightly.'[25]

As the next chapter will show, it's a mistake to believe that 'genes are us' – we are not solely determined by our genes. But something does seem to depend on our genome being common – in both senses – 'common as muck' (unenhanced) and 'common to all'. Even if we each had an individual right to alter our own personal genome, that wouldn't necessarily give us a right to do so for the entire species: 'While the right to personal identity may justify a valid interest in the modification of one's individual genome, the collective right to identity defends a global interest in the preservation of the human genome.'[26]

What social solidarity we possess in an individualistic, market society could be put at risk by the emergence of 'transhumans', according to the American philosopher Michael Sandel:

> *The natural talents that enable the successful to flourish are not their own doing but, rather, their good fortune, a result of the genetic lottery. If our genetic endowments are gifts, rather than achievements for which we can claim credit, it is a mistake and a conceit to assume that we are entitled to the full measure of the bounty they reap in a market economy. We therefore have an obligation to share this bounty with those who, through no fault of their own, lack comparable gifts.*[27]

Sandel's world-view is a long way away from the notion that we have an obligation to be 'the best me I can possibly be': the 'enhancement' equivalent of producing 'the best children that we can'. Instead of that individualistic slant, Sandel's position focuses on social justice. Of course it can be argued that the wealthy already enjoy superior health over the poor, on both the national and the global level. But the fact that there *is* injustice doesn't entail the conclusion that there *should* be *additional* injustice – because both enhancement technologies and techniques for manipulating or testing embryos are of course likely to be monopolized by wealthy individuals and wealthy countries. Then it really would be a world of designer babies, transhumans and lesser mortals.

Are genes us?

While female reproductive biology is being demystified into egg markets and pregnancy outsourcing, genetics has been enjoying a heightened mystique. In 1995 Dorothy Nelkin and Susan Lindee published a book called *The DNA Mystique: The Gene as Cultural Icon*, in which they wrote:

> *Just as the Christian soul has provided an archetypal concept through which to understand the person and the continuity of the self, so DNA appears in popular culture as a soul-like entity, a holy and immortal relic ... It is the essential entity – the location of the true self – in the narratives of biological determinism.*[1]

Genetic determinism is just one variety of the 'biological determinism' that Nelkin and Lindee identify. Other variants include the belief that race or sex determines destiny: that people of colour are suited only to menial work, or that women should confine their career ambitions to motherhood or the nurturing professions. Those views still persist, of course, but thankfully aren't as respectable as they used to be. Genetic determinism, however, appears to be going strong.

There's certainly a persistent popular and media tendency to define personal identity as genetically determined. Candidates for the Least Scientifically Plausible Gene award – if one existed – might include recently 'discovered' genes for getting into debt,[2] becoming a ruthless dictator[3] and voting regularly in elections.[4] But actually, all three of these correlations have at least some minimal basis in science, unlike the entirely speculative 'American exceptionalism gene' posited by the American conservative columnist Michael Medved, who has written:

> In today's ruthlessly competitive international economy, the United States may benefit from a potent but unheralded advantage: the aggressive edge sustained by the inherited power of American DNA.[5]

Medved goes on to argue that there was probably a genetic component to American immigrants' deliberate choice to shoulder the risks of the Atlantic voyage, which still stands their descendants in good stead. He doesn't wince at the contrast with those subjected against their will to the evils of the Middle Passage across the Atlantic from Africa – the enslaved ancestors of today's African Americans:

> The idea of a distinctive, unifying, risk-taking American DNA might also help to explain our most persistent and painful racial divide – between the progeny of every immigrant nationality that chose to come here, and the one significant group that exercised no choice in making their journey to the US. Nothing in the horrific ordeal of African slaves, seized from their homes against their will, reflected a genetic predisposition to risk-taking, or any sort of self-selection based on personality traits. Among contemporary African-Americans, however, this very different historical background exerts a less decisive influence, because of vast waves of post-slavery black immigration.

In the last sentence Medved does concede that modern African Americans might now have a somewhat better genetic propensity for risk-taking, because their original African blood has been diluted, but that's still a 'genes are us' style of argument. It's just that the genes have changed, not that he's had second thoughts about genetics being the basis of identity and behaviour.

You don't have to think too long and hard before you realize the status-quo-is-inevitable implications of this position for racial justice. But in case your genetic background leaves you too lethargic to work it out, Medved goes on to do it for you, claiming that President Barack Obama's 'desire to impose a European-style welfare state and a command-and-control economy not only contradicts our proudest political and economic traditions, but the new revelations about American DNA suggest that such ill-starred schemes may go against our very nature.'

Critics have called Medved's stance just another example of 'American exceptionalism', the founding myth that regards America as different from all other countries – and as implicitly superior. In this instance, 'genetic exceptionalism' – the idea that there's something deeper and more special about genetics than lesser branches of science – is used to back up American exceptionalism. (Of course, other countries, including England, China, France and Iceland, also subscribe to their own versions of exceptionalism. It's the exceptional country that doesn't believe itself exceptional.)

So Medved's claim goes beyond the already ambitious idea that our deepest individual identities are determined by our genes – that 'genes are us', in the useful phrase developed by the British bioethicist Ruth Chadwick.[6] Indeed, he stretches that problematic claim well beyond its snapping point by arguing that the entire USA as a nation is defined by its DNA, and that one of the most disadvantaged ethnic subgroups, African Americans, is inevitably going to be disadvantaged. (He doesn't mention another disadvantaged group, First Nation peoples, but presumably he thinks they also lack 'initiative genes' because they just stayed put.[7])

▲ The Pilgrims land at Plymouth Rock, bringing their supposedly initiative-laden DNA with them.

Even though I'm presumably chock-full of go-getting genes thanks to my *Mayflower* ancestors, I find Medved's argument entirely implausible. His form of genetic determinism – the concept that our most important behaviours are down to our genes – could be seen as just another form of blaming the victim. But genetic determinism is also internally incoherent, as well as scientifically implausible, because it contradicts itself about initiative and free will. This is a common flaw in that familiar assertion about how science – usually neuroscience but sometimes also genetics – has conclusively demonstrated that free will is just an illusion. For example, the US plant scientist Anthony Cashmore recently declared:

> It is often suggested that individuals are free to choose and modify their environment and that, in this respect, they control their destiny. This argument misses the simple but crucial point that any action, as 'free' as it may appear, simply reflects the genetics of the organism and the environmental history, right up to some fraction of a microsecond before any action.[8]

But Cashmore misses an even simpler and more crucial point: he presumably believes we're free to make up our minds whether to deny free will. Why else is he bothering to try to convince us?

Cashmore's argument fails to take account of the important distinction made by the eighteenth-century Scottish philosopher David Hume: while all human actions are caused, they could still be free. Even if genetics were eventually able to demonstrate that all action is genetically caused – which seems extremely unlikely – that would not be enough to disprove free will. Indeed, it takes an act of free will to assert that all human action is externally determined, and so all determinism – including genetic determinism – is self-contradictory.

Ironically, 'get up and go', American or otherwise, is the first casualty of genetic determinism. If you really believe it's all down to the genes, there's no room at all for individual initiative or autonomy. Those, too, are merely the products of your genes, so you don't deserve any credit for them. Genes command, in this view, and we obey. In one well-known variant of this thesis, Richard Dawkins's *The Selfish Gene*,[9] not only do 'we' obey our genes: our genes obey the dictates of evolutionary success. There's not much left of 'us' at all, in fact: our personal identity has been obliterated. So why do so many people apparently want to believe that genes are indeed us?

Decoding genetic terminology

Definitions of terms in the text can be found below.[10]

Alleles Alternative forms of a particular gene.

DNA (deoxyribonucleic acid) The chemical that encodes genetic information.

Epigenetics/epigenomics From the Greek *epi* (above), the science of how genes are expressed to create different patterns of formation during embryo development and through their interaction with their environment.

Gene An inherited instruction that tells the body how to make proteins.

Genome The entire genetic information of any living thing.

Genotype The specific DNA sequence present at a particular chromosome location ('what you can't see').

Human Genome Project The international effort to decode the entire genetic information of a human being.

Phenotype The emergent manifestation of a particular gene variant ('what you can see').

Recessive A pattern of inheritance caused by a gene that only has an effect if two copies are inherited, one from each parent. The parents themselves will not manifest the condition if they each carry only one adverse allele of the gene, so they may be unaware of their carrier status.

Are genes us?

First, not all behavioural genetic correlations are as spurious as the one put forward by Medved. For example, in May 2011 the *Journal of Human Genetics* published a study that links your satisfaction with life to the type of 5-HTT gene found in your genome. This gene encodes a transporter for the brain chemical serotonin, which has been shown to be involved in depression and mood. Of the 2,574 Americans surveyed, 69 per cent of those who had two copies of the long version of the gene were satisfied or very satisfied with life, compared with only 38 per cent of those with two copies of the short allele.[11] Another recent study showed some genetic susceptibility to starting smoking and to difficulty in stopping.[12] Such genetic links to behaviour and character may be exaggerated by the media, but some have a genuine basis in evidence.

A second factor behind the popularity of the 'genes are us' view is the manner in which genetic research has been systematically 'sold' over the past 40 years – since *Time* magazine published an issue on 'the new genetics' in 1971. Genes have been hailed as the building blocks of biology, the blueprint for how to build a human and 'the language of life',[13] in Francis Collins's phrase. It seems clear from their own accounts that many of the Human Genome Project researchers, such as Collins and the British scientist John Sulston, genuinely believed that sequencing the entire human genome would bring tremendous benefits to medical research in a comparatively short time.[14] When the first draft of the human genome was sequenced in 2000, an editor at the respected science journal *Nature* even predicted that by the end of the twenty-first century:

> *Genomics will allow us to alter entire organisms out of all recognition, to suit our needs and tastes ... [and] will allow us to fashion the human form into any conceivable shape. We will have extra limbs, if we want them – maybe even wings to fly.*[15]

Whether such overheated claims were encouraged by commercial interests is an interesting speculation, which has been studied systematically by the British genetics researcher and activist Helen Wallace. In a painstaking and superbly documented report, Wallace has uncovered new evidence of tobacco and food industry involvement in what she views as the murky origins of the Human Genome Project. It was in the tobacco industry's interest, she argues, to isolate some individuals as genetically prone to lung cancer, so that anyone with the genetic all-clear could continue to puff cheerfully on. (Unfortunately for the industry, a large twin study done in 1995 actually found no genetic basis for lung cancer.)

According to Wallace's report, the president of the Council for Tobacco Research boasted before Congress that his organization had funded a thousand researchers, to the tune of nearly $225 million, for work in identifying familial cancers. That programme was closed down in 1999, but British American Tobacco informed the House of Commons in 2000 that it was still funding research on genetic susceptibility to lung disease. As Wallace declares forcefully:

> *Claims that human genome sequencing will be useful to predict who develops common diseases are false and originate from spurious findings published by tobacco-funded scientists. The food and pharmaceutical industries have also promoted false claims that human genome sequencing will predict big killer diseases, in an effort to expand the market for healthcare products to large numbers of healthy people and to confuse people about the role of unhealthy processed foods in hypertension, type 2 diabetes and obesity.[16]*

A third reason why the 'genes are us' view continues to be popular is that people probably don't realize the comparative lack of impact on medical treatment that genetics has actually had. 'Indeed, after ten years of effort [since the

Human Genome Project] geneticists are almost back to square one in knowing where to look for the roots of common disease.'[17] For example, a recent study of 101 genetic variants that had been statistically linked to heart disease in different genome-wide studies showed no value at all in forecasting the disease for a group of 22,000 white US women over a ten-year period.[18] The old-fashioned method, taking a family history, was actually found to be more informative. Yet Francis Collins predicted that genetic diagnosis of common diseases would be routine by 2010 and that cures would follow five years later.

After a few early successes with atypical single-gene conditions such as Huntington's disease, genetic and genomic research has had far less immediate impact on the practice of medicine than was originally hoped. Because major diseases like cancer and cardiovascular illness are common, so too would be their genetic causes, researchers surmised. As it has turned out, however, the common variants explain just a fraction of the genetic risk. But as the American physician and bioethicist Howard Brody slyly remarks, these inconvenient truths have actually been transmogrified into a selling point to research funders:

> How quickly the leading genomic scientists in the United States changed their tune from how much would be known once we completed mapping the genome (so keep that generous funding coming our way) to how little was actually known once we completed mapping the genome (so keep that generous funding coming our way).[19]

Take cystic fibrosis, the most common recessive genetic condition in white people: about one in every 25 individuals of European descent carries a mutated copy. While the mutation has been successfully identified, no one has so far developed any therapies targeted at the CFTR protein that it expresses. The 'cystic fibrosis gene' itself has become the basis of much further genomic research, some of it potentially productive. However, as some commentators in the field themselves admit, sequencing the human genome has done a lot for science but not much for medicine. The researcher Jack Riordan, who collaborated in the original sequencing of the cystic fibrosis gene, puts it this way:

> The disease has contributed much more to science than science has contributed to the disease ... [Developing a cure for cystic fibrosis] is not like going to the moon – it's going to Mars.[20]

One way in which genetic science has developed prodigiously in recent years, but which in some senses has delayed its impact on medicine, is progress in epigenetics. For example, researchers are now investigating not just the genetic basis of cancer but also its epigenetic basis. All cells in the human body carry the same complement of about 25,000 genes, but those genes 'come out' very differently in, say, the heart and the retina. What factors activate or silence the way genes are expressed? This could be tremendously important in establishing what turns having a genetic propensity to cancer into having the disease itself.

So epigenetics has moved well beyond the straightforward association of 'one gene, one trait', or the notion that genotype accounts for every aspect of phenotype. Likewise, epigenetics could undermine the oversimplified notion that 'genes are us', and that can't be a bad thing. As one leading scientist, Edith Heard, wrote in a ten-years-on review of what's been achieved since the sequencing of the human genome, 'Epigenetics may provide hope that we are more than just the sequence of our genes – and that our destiny, and that of our children, can be shaped to some extent by our lifestyle and environment.'[21]

Why everything you learned about how acquired changes can't be inherited is not wrong – but not quite right either

I certainly had it drummed into my head during high-school biology that changes made to any organism during its lifetime couldn't be passed on to its descendants. If you cut a rat's tail, for example, the rat pups it produces will still be born with tails. And I also learned about how backward the contrary view was – associated with the eighteenth-century scientist Lamarck and the pseudoscience encouraged by Stalin.

Strange thing, the forward march of science. Epigenetics has now discovered a mechanism called 'imprinting', which undermines that straightforward view, though emphatically without proving Lamarck (or Stalin) right either. As the British emeritus professor of molecular embryology Marilyn Monk has explained:

With the discovery of imprinting, the differential expression of paternally inherited and maternally inherited copies (alleles) of certain genes, it became clear that certain modifications influencing the potential of a gene to be expressed could pass through the germ line – through the egg or the sperm. This is a highly significant discovery, as until imprinting was demonstrated, a major objection to the possibility of Lamarckian inheritance (transgenerational inheritance of acquired characteristics) was the lack of a conceivable molecular mechanism. The passage of epigenetic modifications through the germ line means that the way we live our lives may influence the potential of the genes we pass to our offspring, and imbues upon us a certain longitudinal responsibility for future generations.[22]

So far I've been quite sceptical about some of the founding myths of popular genetics: the genetic mystique, genetic exceptionalism and the scientifically disinterested nature of the Human Genome Project. When genetic determinism is used to justify the underprivileged status of a particular ethnic group, as Medved's argument seems to do, then I think it's particularly suspect. But what if such a group can actually turn genetic determinism to their advantage?

That occurred in Tonga, when an Australian firm, Autogen, announced in November 2000 that it had concluded an agreement with the Tongan Ministry of Health to collect a bank of tissue samples for genomic research into the causes of diabetes.[23] (The Tongan population had an exceptionally high rate of diabetes, about 14 per cent.) In the press release

Autogen remarked that it was attracted to Tonga because of its 'unique population resources'. Isolated populations such as Tonga's are particularly appealing to researchers: when Iceland's deCODE Genetics announced a somewhat similar deal, it made much of Iceland's exceptionally homogeneous population (although that was later disproved).

Although the Tongan public hadn't been informed beforehand, Autogen might have expected little resistance. It was offering quite generous benefits, including research funding for the health ministry, royalties to the Tongan government from any successful discoveries, and the free provision of any new drugs created from them. But what Autogen hadn't counted on was that the Tongan culture did indeed regard their genes as something like the 'holy' substance Nelkin and Lindee discuss. To the Tongans, the benefits on offer from Autogen smacked of trinket exchange.

Led by the charismatic Lopeti Senituli, the Tongan resistance movement successfully insisted that their genetic information shouldn't be the object of a commercial deal, because it was held on trust from their ancestors for their descendants. Strictly speaking, this wasn't quite the same concept as the 'genetic mystique'. It was expressed in terms of the Polynesian and Maori cultural concepts of *ngeia* (dignity of the person)[24] and *tapu* (sanctity of the individual). In a modern updating of these traditional concepts, however, the eminent Maori studies professor Hirini Moko Mead has written of tapu: 'This attribute is inherited from the Maori parent and comes with the genes.'[25]

So what's the difference between Mead's claim – 'it comes with the genes' – and the form of genetic determinism that I've done my best to discredit? One key difference is that the indigenous view is inherently communitarian, not individualistic. It is used not to glorify individual initiative or to condemn communal healthcare

▲ The Ojibwe novelist Louise Erdrich, whose tribe cautioned her that DNA samples from her body do not belong to her alone.

provision, as Medved does, but to stake a claim against commercialization and possible exploitation of the ethnic group, in the name of solidarity.

Of course genetic heritage, in this communitarian view, also puts restrictions on what individuals are allowed to do with their 'own' genome. The Ojibwe (Chippewa) novelist Louise Erdrich tells a story of how she was contemplating taking a personal genetic test, of the sort you can now purchase over the Internet by providing a spit sample and a certain amount of cash. 'But when I asked my extended family about this – and I did go to everyone – I was told, "It's not yours to give, Louise." '

This isn't just a rarefied view that pertains only to indigenous populations: it raises some of the most profound problems about genetic testing – such as whether a diagnosis with implications for other family members should be revealed, if the individual patient wants to keep it quiet. Do extended family members also have a right to know?[26] This is the same line of logic as Erdrich's family was pressing on her (so that she decided not to take the test). They viewed genetic information as a core part of their identity, but didn't regard it as individually 'owned'.

In the next chapter we'll look in more detail at who 'owns' genetic information in a rather different context: genetic patenting, a very live and potentially lucrative issue in today's bioethics.

Could you patent the sun?

Whether or not genes really are the 'language of life', you might be surprised to learn that the language has been substantially privatized. By 2005, the number of patented human genes totalled 4,270, almost one-fifth of the human genome.[1] By now the number is probably substantially higher. These patents are typically exclusive, so that research on the gene can take place only with the permission of the patentholder.

If you were a speaker of 'Genetiski' as a language, you would effectively be barred from using every fifth word as you constructed a sentence.

Let's put my last sentence in similar terms: 'If you were a CENSORED of "Genetiski" as a CENSORED, you would effectively be CENSORED from using every fifth CENSORED as you constructed a CENSORED." You can see that this would create a communications logjam, rather like the effect of genetic patents on researchers as they attempt to carry out their scientific business.

But we saw in Chapter 2 that the common law traditionally regarded tissue taken from the body as *res nullius*, 'no one's thing'. If tissue once taken from the body belongs to no one, then how can it be the subject of any kind of property – including intellectual property?

Controversial precedents – with which many legal experts disagree – have been set by a case in which a genetically engineered bacterium was successfully patented[2] and by the Harvard 'onco-mouse', a genetically engineered mouse developed and patented for cancer research.[3]

But the mouse and the bacterium are living things. How can you take out a patent on life?

Even if you could – on the grounds that we can own pet mice or tubs of probiotic yoghurt – shouldn't we draw the line at human genes and human life? In non-slave societies, we can't own another human being. So how can anyone own another human being's genes? Yet the argument that genetic patents are tantamount to slavery has failed to convince judges.[4]

The patenting of human genes is a tough test for those who think that we should just let science do whatever it wants, and that clinical care will automatically benefit. The effect of gene patents is not necessarily to improve science, but often to impede both research and clinical care.[5]

For example, the cancer drug Herceptin®, which acts on the *HER2* (human epidermal growth factor receptor 2) gene, can benefit women whose breast cancers are positive for the *HER2* gene, which makes the tumour more aggressive. But the biotechnology firm Genentech holds patents not only on the drug – which you might think is fair enough – but on the gene itself. Any researcher or drug company wanting to develop an alternative, cheaper drug must obtain permission from Genentech or risk being sued for patent infringement.[6] Despite its efficacy, many NHS hospital trusts had to restrict the use of Herceptin because the price was too high: no other company could produce a cheaper drug without access to the gene.

Anyone who believes that we should let biomedicine do what it wants, so that unimpeded progress can flourish, is faced with a troublesome contradiction: scientists have conflicting interests. Rival scientists want open access to patented genes, but the structure of modern biomedical research depends on the 'promissory capital' embodied in patents and held by universities or their private funders.[7] (Very often patents are the most valuable component of a start-up firm's portfolio.) Clinicians want access to evidence-based therapies for their patients, deriving from other people's research. But many clinicians are also academic research scientists, who may well be under pressure from their universities to take out patents on any discoveries they make, and thus restrict access. It's a paradox of which Zeno could be proud.

So it's impossible for policy-makers simply to give science what it wants, because science is not united or unitary. We have to make some difficult choices, and those choices have to be reasoned out in legal and philosophical debates. Sometimes those debates get very hot indeed, particularly when very substantial commercial interests are involved. Things have moved on – or backwards – considerably from the days when Jonas Salk replied to a newscaster's query about who would own the patent on his polio vaccine: 'Well, the people, I would say. There is no patent. Could you patent the sun?'

A crucial current case about restrictive patenting has concerned diagnostic tests on two genes implicated in some breast cancers, *BRCA1* and *BRCA2*. Women with the 'wrong' version of these genes have a heightened risk of developing breast cancer (up to 85 per cent, compared with the normal 12 per cent, although the genes account for only a minority of breast cancers). These women also run a greater risk of ovarian cancer.

The US firm Myriad Genetics, which, along with the University of Utah, holds patents on both genes, has charged fees of around $3,000 to American women who thought they might be at familial risk and who therefore wanted a diagnostic test. One such patient was Lisbeth Ceriani, a 43-year-old woman with breast cancer whose doctors recommended that she should be tested for the unfavourable alleles of the BRCA1 and BRCA2 genes. Myriad did not accept her insurance, and Ceriani could not afford to pay for the test. So she remained ignorant, as did her physicians – with possible adverse ramifications for her clinical care.

The legal basis in Myriad's defence was that what is being patented is not the gene in my or your body, but a 'cloned' version of the gene created in a laboratory. Rather than a 'patent on life', the company said, they are patenting something more like a synthetic chemical. So, in a sense, 'no one's thing' remains applicable to the actual gene in my body or your body: Myriad doesn't own any part of our individual genomes.

Yet that seems very paradoxical: if the patent isn't on the actual gene in my body, how can Myriad rightfully charge a fee to diagnose whether I have the faulty version of the gene? Isn't this a sort of abuse of my human rights? That was part of the argument put by an alliance of medical associations, the American Civil Liberties Union, and patients. They succeeded in overturning most (but not all)of the Myriad patents in a Federal district court in March 2010.[8]

The Association for Molecular Pathology, the American College of Medical Genetics, the American Society for Clinical Pathology and the College of American Pathologists – together with individual patients, including Lisbeth Ceriani – were all part of the line-up against the US Patent and Trade Office, Myriad and the University of Utah. (The involvement of the medical associations shows how pernicious and simplistic it is to accuse critics of commercial biotechnology of being opposed to medical progress.) Over 150 laboratories attested that they had been blocked in their scientific research by 'cease and desist' orders from Myriad. But the district court judgment was overturned on appeal in July 2011. In March 2012 the Supreme Court sent the case back to the appeal court, which will decide in July 2012.

The justification for genetic patents, relied on by Myriad and other firms or universities, is the idea that what is being patented is not a discovery of something pre-existing in nature, but rather an invention. Because an invention represents labour, skill and perhaps capital investment, this logic goes, it can rightfully be patented. As the US academic lawyer Rebecca Eisenberg puts it:

> One cannot get a patent on a DNA sequence that would be infringed by someone who lives in a state of nature on Walden Pond, whose DNA continues to do the same thing it has done for generations in nature. But one can get a patent on DNA sequences in forms that only exist through the intervention of modern technology.[9]

This argument rests on the philosophical basis put forward for property by the English philosopher John

Locke (1632–1704): that it depends on 'mixing one's labour' with materials such as land and tools. Labour, skill and concerted effort are the sources of an object's value, in this view, and of the right to it. Locke distinguishes much more carefully than some of his modern followers between having a property in the body itself – which he denies – and having a rightful property in the results of bodily labour, which he recognizes as an expression of human agency.[10] In that form, it's actually not just a justification for the wealthy to hang on to their wealth, but a progressive argument with a great deal of potential for exploited groups, as I've argued elsewhere in relation to women's property in eggs 'donated' for research.[11]

Locke's work is also the unexpected basis for the labour theory of value and the related concept of surplus value, crucial to the analysis of Karl Marx (1818–83). In Marx, exploitation can objectively be said to exist when the surplus value, the difference between the cost of production and the amount paid to labourers, is seized by the owners of the means of production. That analysis can and has been applied to 'biocapitalism', which is generating its own exploited classes, in a Marxist view.

But Locke's labour-mixing metaphor was intended to apply to tilling a vegetable patch or planting a Worcester Pearmain tree, and then laying claim to the turnips and apples. Marx, likewise, was concerned with a different, earlier mode of production: industrial production. It takes quite a bit of nipping and tucking to tailor these analyses snugly to modern biotechnology. In the case involving the genetically modified bacterium, Diamond *v*

Chakrabarty, the court nevertheless did follow Lockean reasoning when it ruled that:

> *Anything under the sun made by man can be patented.*

So the more 'man-made' – the further from the natural state and the more labour involved – the better, as far as the claim to have invented rather than discovered the patented form of the gene is concerned.

The *Chakrabarty* court accepted that patents are a rightful reward for research costs, labour and know-how. In addition to this Lockean rationale, the judges also adduced the utilitarian argument that without patent protections, research and clinical care would shudder to a halt – although that turned out to be doubtful in the Herceptin case.

But how much work does it take to justify a property claim in the form of a patent? The researcher in the case, Chakrabarty, admitted that he simply 'shuffled the genes' of an existing bacterium, and that no great effort or skill was required. Similarly, most genetic patents are now taken out after routine procedures involving large-scale computer arrays, not radically new inventions.

Dr Frankenstein could well have argued that he had worked long and hard by flickering candlelight in his dank laboratory, exposed to risky lightning strikes through the open roof and hampered by a shortage of assistants with unbroken necks. Arguably, he would have had a stronger claim to a patent on his monster than Myriad does on the *BRCA1* gene.

Unlike the *Chakrabarty* court, the district court judge in the *Myriad* case, Judge Sweet, was not satisfied that sufficient invention had taken place to justify a patent. He ruled against Myriad on the question of invention versus discovery, saying: 'The claimed isolated DNA is not markedly different from native DNA as it exists in nature; it constitutes unpatentable subject matter'.

This sceptical view was also upheld in a supporting brief from the US Justice Department, which remarked that Myriad had also claimed patents on genes directly, without the labour of isolating or cloning them. The Justice Department specifically distinguished human DNA from a genetically engineered bacterium. In a statement which the activist blog FierceBiotech described as having 'dropped a bombshell on the biotech industry', the Justice Department brief declared:

> Unlike the genetically engineered micro-organism in Chakrabarty, the unique chain of chemical base pairs that induce a human cell to express a BRCA protein is not a 'human-made invention' . . . Common sense would suggest that a product of nature is not transformed into a human-made invention merely by isolating it.[12]

Though Judge Sweet didn't stress the fact, it wasn't even Myriad that made the original 'discovery', but the publicly funded charity Cancer Research UK. That raises important issues about public investment versus private profit. Both scientific knowledge and the genome itself can be construed as public goods; genes are in a sense the common heritage of humanity, according to the 1997

UNESCO Universal Declaration on the Human Genome and Human Rights.[13]

Just as Louise Erdrich's DNA wasn't hers to give, in the view of other members of her Ojibwe people, so it could be said that our human DNA isn't anyone's to take. When firms are allowed to take out patents on this 'genetic commons', what does that say about our attitude towards nature? During the *Myriad* case it began to appear possible that the whole of nature – both organic and inorganic – might be subject to a titanic patent grab.

How about the chemical building blocks of the universe? Yes, those too...

As the *Biopolitical Times* reported in its coverage of day one of the Myriad appeal on 6 April 2011:

Perhaps the most striking moment of the day came when Myriad's lawyer unaccountably admitted that under the company's legal theory, elements of the periodic table are in fact eligible for patents. Pressing the logic of this claim, the Solicitor General pointed out that lithium, a naturally occurring element that does not exist in isolated form in nature, was first isolated by a chemist in 1818, but that no one has ever claimed it is patentable. In response to questioning by the judges, Myriad's lawyer said pretty much outright that, yes, he does believe that isolated lithium is patentable material. In his defence, he did note that he was an English major, not a scientist.

The US law professor James Boyle believes that we are living through a transition as momentous as the enclosure of the English agricultural common lands in the eighteenth century or the Scottish Highland clearances in the nineteenth. Enclosure of the agricultural commons provided the capital for new industries and the labour too, by dispossessing the rural poor from their land and sending them flocking into the cities, where they were vulnerably available to become the exploited urban poor of the Industrial Revolution. Similarly, patenting as a form of enclosure provides promissory capital for start-up biotechnology firms, while converting what was formerly a common good – the human genome – into a new form of private property.[14]

Like the agricultural enclosures, the 'great genome grab' – unless kept in check by actions like the Myriad case or the Tongan resistance discussed in Chapter 4 – would transfer wealth from the public domain of the 'genetic commons' to biotechnology venture capital. In the Herceptin example, we've already seen the defects in one justification for this massive appropriation – that research and clinical care will automatically benefit, because patents provide an incentive for scientific progress. What about another commonly heard argument in favour of privatization of public resources – the 'tragedy of the commons'?[15]

According to this view, communal property is inherently prone to abuse, because everyone who has common rights in an object is tempted to overuse it. The defenders of land enclosures explicitly argued that transferring inefficiently managed or overgrazed common land into single ownership and clearing off the tenants would improve agriculture and increase food production for the country as a whole. As the aristocratic landowner in the Scottish novelist Neil Gunn's novel *Butcher's Broom* puts it, in conversing with his wife about turfing out her tenants and turning the land over to sheep:

> *Now on your estates you know the conditions. Apart from whether the people are ignorant and slothful and bestial in their habitations, we do know that they live in poverty . . . . Well, here's a scheme that is going to use the land in the only way it profitably can be used. You will benefit largely, but not more than the country, for the estate will now export huge quantities of wool and mutton. Ultimately what benefits the country as a whole benefits the people as a whole.[16]*

But even if it were granted that the enclosures increased food production – for those who could afford it – it's hard to see how the parallel works. The argument in favour of enclosing the common lands was that they were being wastefully used, but how could any of us 'overuse' the human genome?

Indeed, if there is truth in the allegations by the 150 laboratories who say they were blocked from doing research on the *BRCA1* and *BRCA2* genes by threats of legal action from Myriad, then what is at stake is actually the tragedy of the anti-commons: a private right to prevent others from using property of mutual interest, resulting in underuse of the monopolized resource. The Herceptin case is another example. In both cases, patients have been blocked from using the diagnostic tests or treatments of which they have dire need, because they or their national health systems can't afford the monopoly price.

Like the laird in *Butcher's Broom*, patent holders have been allowed to claim that the patent system is good for the collective, as well as in their individual interests. As I wrote about the more inflated claims of the patenteers in *Body Shopping*:

> What's good for Myriad Genetics is good for the world, in this view. Even General Motors more modestly claimed that what was good for it, was good for the nation.[17]

But the high tide of patent monopolies may have passed. Although Myriad won its case at the appeal court level, it's by no means certain what the final outcome will be, as I write. And there are also a number of other recent decisions restricting the pell-mell growth of patenting.

In March 2011, just before the *Myriad* case appeal began to be heard, the European Court of Justice was asked to give an opinion on whether embryonic stem cell lines could be patented, given their origins in a human embryo. The challenge, mounted by Greenpeace against a patent on a technique to generate nerve cells from embryonic tissue, resulted in a preliminary judgment in Greenpeace's favour, although, like the *Myriad* case, this one will probably run and run. That surprising outcome raises the issue of whether there is something specially controversial about embryonic stem cells. But the answer to that question will have to wait until the next chapter.

6

Snowflakes, techno-coolies and the Tooth Fairy: some wonders of stem cell science

ALL THAT MATTERS

Stem cell research makes similar promises to enhancement technology, but not just about whether we can improve our intelligence, appearance and strength. Those goals are comparatively trivial next to the hopes invested in stem cell research – curing blindness caused by retinal atrophy, for example, or alleviating spinal paralysis. The first trial of stem cell therapy for the latter was started in 2011,[1] although called off that November. The former has got as far as a small-scale trial to establish clinical safety. This puts stem cell research slightly ahead of the more speculative forms of enhancement technology, although not by as much as the papers tell us.

Many people are under the sadly mistaken impression that stem cell cures are just around the corner. But as a regenerative medicine expert told me recently, 'The challenges of translating stem cell research to the actual regenerative medicine component are largely the same as would have been identified five years ago.' The US bioethicist Art Caplan thinks vested interests lie behind the widespread faith in imminent stem cell cures:

> *Embryonic stem-cell research was completely overhyped, in terms of its promise. And people knew it at the time . . . [T]his notion that people would be out of their wheelchairs within a year if we could just get embryonic stem-cell research funded was just ludicrous . . . The scientists had to have known that. Who has ever delivered a cure in a year from something that's basically a dish? . . . But the politics of that issue were abortion politics, meaning that one side had as a principle, 'Don't kill.' The other side had as a principle, 'You've got to cure.' And that escalated*

> *the rhetoric. So I think the science got hyped in response to the politics.[2]*

Caplan is referring particularly to the US context, where in 2001 the religious right succeeded in banning federal funding for stem cell research. The stem cell issue was particularly polarized and bitter there, and still is: although President Obama rescinded his predecessor's ban on funding, the issue is still being dragged out in the courts. In Britain the religious divide has been less cavernous, but similar exaggeration occurred during the 'cybrid' brawl of 2008 (see Chapter 1). Opponents of cybrids were portrayed as opposing all stem cell research and as being rabidly 'religious', neither of which was true.

So a certain level of misinformation and 'hype' has typified stem cell debates. Certainly when Hwang Woo Suk published his false claims to have invented a stem cell 'personal repair kit' (see Chapter 1), a media firestorm smoked out most sceptical opposition – with a few stalwart exceptions. One of them, the US analyst and activist Marcy Darnovsky, was concerned not just about the feasibility of Hwang's somatic cell nuclear transfer (SCNT) technique, or even the risks it posed to women egg donors, but also the social justice of it. As she wrote:

> *But it's important to recognize that the 'personal repair kit' scenario is at best very remote, and will likely always be unrealistic. That's because medical treatments based on SCNT would be extraordinarily expensive. An article in the* Proceedings of the National Academy of Sciences, *for example, estimates a cost of $100,000 to $200,000 just for the preliminary costs of deriving a cell line for a single*

ALL THAT MATTERS: BIOETHICS

patient this way. So SCNT-based treatments, even if they turn out to be technically feasible, would almost certainly be a kind of 'designer medicine,' out of reach except for the very wealthy.[3]

We haven't heard so much about these justice-based doubts, however, as about the rights and wrongs of using human embryos in some forms of stem cell research. Hwang's SCNT technique didn't use surplus embryos from in vitro fertilization (IVF); rather, it demanded a copious supply of human eggs. Many commentators at the time thought that was an ethical improvement over

▲ Hwang Woo Suk, who used over 2,200 human eggs in stem cell research that was later found to be false.

the existing embryonic stem cell technique developed by James Thompson's team, which did rely on 'spare' embryos (see box). Even after the revelations about Hwang's exploitation of his donors, the British medical lawyer Emily Jackson, for example, declared that those issues had been overplayed, and that the real scandal in the Hwang case was the waste of embryos.[4]

Not all stem cells are created equal

A stem cell is an undifferentiated cell that can give rise to daughter cells diversifying into a variety of other cell types. Many people think that all stem cell research is basically the same, but that's definitely not the case.

Embryonic stem cells (ESCs) These cells, first isolated and cultured by the Thompson team at the University of Wisconsin in 1998, are derived from the inner cell mass of a human blastocyst (a very early stage in the fertilized embryo).[5] 'Immortal' cell lines can be created from them, but so far they've yielded no therapies, unlike adult stem cells and haematopoietic stem cells (see below). Embryonic stem cell lines also have the disadvantage of not being matched to the patient, possibly triggering immune rejection. The next two types of stem cell lines were developed to avoid that problem, since, theoretically at least, they could be developed from a body cell donated by the patient.

Induced pluripotent stem cells (IPSCs) First developed in 2007 by the Yamanaka team in Japan,[6] this type of stem cell does not occur in nature but can be induced from adult tissue by epigenetic reprogramming. (If you remember that all cells in the body have the same genes, it should be possible to turn on 'silenced' genes, transforming, say, a nerve cell into a muscle cell.) At first it was feared that this process might be carcinogenic, but that problem seems to have been conquered. Some high-profile researchers – such as Ian Wilmut, creator of 'Dolly', the cloned sheep – largely switched over to IPSCs as the way of the future. But in May 2011 researchers announced the surprising and discouraging result that IPSCs made from mouse skin cells were rejected by the mice's immune systems.[7] Even if this also turns out to be true in humans – which isn't definite – IPSCs will still be crucial for modelling 'disease in a dish'.

Somatic cell nuclear transfer See the description in Chapter 1.

Fetal stem cells Taken from fetal germinal ridge tissue after early-stage abortion, these cells now appear most likely to be confined to neurological research.

Adult stem cells These cells have the capacity to become many types of body tissue but are not as flexible as embryonic stem cells, because they are more differentiated. However, precisely because ESCs are so 'plastic', they can also form tumours. Other researchers are therefore concentrating on adult stem cells, with some promising results.[8] An impressive recent success was the successful transplantation of a human trachea (windpipe) using a donor trachea covered with adult stem cells from the transplant recipient, cleverly avoiding immune rejection.[9]

Umbilical cord blood/bone marrow (haematopoietic stem cells) These two types of 'blood-making' (haematopoietic) cells may have very different origins, but they've been found useful in similar conditions, such as leukaemia. Bone marrow cells are typically (but not always) taken from adult bone marrow and are less 'plastic' than cord blood, but fully tissue-matched bone marrow transplants still give the best results for patients with leukaemia.[10] Umbilical cord blood is frequently taken between the second stage of childbirth (delivery of the baby) and third stage (delivery of the placenta) to maximize collection, although it is less risky to take it after the placenta has been delivered. It can be either autologous (the patient's own) or allogeneic (from someone else). The current medical consensus is that while the value of public (allogeneic) banking is well established, autologous private banking has a worse record, including a higher risk of relapse for patients with leukaemia.[11]

Often critics of one particular form of research are accused of endangering all stem cell research by their opposition. But the techniques are so very different in terms of their methods and materials that it's quite legitimate to favour some but not others.

Personally, I have few ethical qualms about induced pluripotent, bone marrow and adult stem cells; indeed, I think they're positively a good thing. I have very grave doubts about SCNT and about privately banked autologous cord blood techniques, both because (along with many others) I doubt that the science is productive and because I think they exploit women's altruism while commercializing their tissue. Publicly banked cord blood is clinically more effective, but the line between public and private is getting blurrier as public banks sell cord blood on international markets.[12] Somewhere in between, for me, come fetal and embryonic research: I don't think the embryo or fetus is a person, but I do think there are complex issues about how a woman views her relationship to the developing embryo or fetus.

The common law says that personhood does not begin until birth, although abortion legislation in the UK and elsewhere does give increasing degrees of protection as the fetus matures. No living human being is killed when IVF results in more embryos than can safely be implanted in the woman's uterus. You can't even say that a potential human being is killed, because that potential will never be fulfilled unless a woman voluntarily accepts having the embryo implanted.

Typically, moderate hormone stimulation treatment results in something between 10 and 15 eggs being produced. If all were viable, and all were successfully fertilized, that would mean 10–15 embryos. That's improbable, but it's quite likely that multiple embryos will be created. If more than one embryo is created, guidelines in the UK and many other countries suggest that only one should be implanted, because of the well-evidenced hazards to both mother and fetuses of multiple pregnancy. The remaining embryos can be frozen for a limited period in some countries, but if the mother doesn't want more children, eventually a decision will have to be made about whether they are to be donated to others, used for research or destroyed. Although I don't usually think that choice is the be-all and end-all of bioethics, I see no reason why a woman shouldn't choose to donate them to research rather than destroy them (although I would also think it perfectly understandable if she preferred not to let them be used by either researchers or another couple).

Even anti-abortionists ought to be able to see the difference between forbidding a woman to interrupt an existing pregnancy and forcing her to start a new pregnancy with more embryos than is medically advisable. If there were seven or eight embryos, say, the likely result would be that the woman's health was endangered and all the embryos died – which a right-to-life advocate would presumably view as a worse outcome than a successful singleton pregnancy. (True, we had 'Octomom', but apart from her, and the similar number born to cartoon characters Apu and Manjula on *The Simpsons*, successful octuplet pregnancies are very rare indeed.)

'Snowflakes': 'adopting' an embryo

The closest the 'anti' brigade has come to recognizing the basic truth that embryos require a woman's womb to grow in before they can become persons is the 'Snowflake' 'adoption' programme for embryos. Set up in 1997, 'Snowflakes' is run by Nightlight Christian Adoptions. Its website describes its aims as:

> *Helping some of the more than 400,000 embryos realize their ultimate purpose – life – while sharing the hope of a child with an infertile couple.*

Snowflake Number 265 was born in April 2011. Only 399,735 to go, provided no more surplus embryos are created in the meanwhile – although they certainly will be.

But let's bypass all the angels-waltzing-on-the-head-of-a-pin arguments about when life begins, and grant opponents of embryonic stem cell research their claim that the embryo is a person – just for argument's sake. That actually puts them in a legal and logical cleft stick: no one can be forced to undergo a medical procedure against their will for another person's sake – not even if it would save the second person's life.

This principle has been established for years: for example, in the US case McFall *v* Shimp,[13] in which a man refused to undergo a life-saving bone marrow transplant for his cousin's benefit, even though he was the only suitable tissue match. If the embryo is a person, then a woman cannot be compelled to undergo its implantation into her uterus, to go through pregnancy and to endure childbirth.

It's quite surprising how often the woman's essential role seems to get left out (as if the artificial uterus were already a reality). Or is it? Perhaps this is just another aspect of the phenomenon I've termed 'the lady vanishes', which was also obvious in the initial euphoria about another form of stem cell research, SCNT. Nevertheless, it's undeniable that the stem cell debates have been dominated by the issue of whether the embryo or fetus is a person.

But even if we deny that the embryo is a person, we might well feel uneasy about equating embryos with other kinds of discarded tissue. The European Patent Court judgment referred to at the end of Chapter 5 turns on the foggy but intuitive notion of the dignity of the discarded embryo. What's the supposed source of that dignity? The court located it in the fact that, if implanted, the embryo could give rise to an entire human being. By contrast, the court held, pluripotent adult stem cells can't form an entire person, only specific tissues. But I'm not quite sure why that should make a moral difference.

Yet even though I don't think any human being is killed in embryonic stem cell research, I can see the merit in arguments that embryonic tissue isn't just another consumerized object of commerce. However, I take a broader view of all human tissue as 'uncommodifiable', whether or not it could form a complete person (although I recognize that, pushed to the extreme, that argument would forbid selling hair for a wig). Just as a single human gene in its own right can't be patented – only a cloned version – so human embryonic tissue deserves legal protection. The French, whose bioethics legislation is based on the idea that human tissue shouldn't be commodified, use that kind of argument all the time.

> *L'embryon n'est ni une personne ni un matériau.* ('The embryo is neither a person nor a raw material.')

> *Hervé Mariton, French Assembly deputy, during debates in May 2011 on reframing bioethics laws in France*

But that doesn't tell us what kind of protection is appropriate – in particular, whether other forms of tissue used to make stem cells deserve more, less or the same level of protection. We saw in Chapter 1 that the human eggs used in Hwang Woo Suk's SCNT research barely registered on the ethical radar of researchers. There's been a tendency to assume that any form of stem cell research that doesn't involve human embryos must be ethically permissible. If there is any form of research that is likely to get the thumbs-up from most bioethics experts, it's induced pluripotent stem cells, but even there doubts have been raised on ethical as well as practical grounds.[14]

In order to decide what kinds of regulation and protection are appropriate, we first have to be accurate about what's going on and who is affected. And we also need to recognize that even though embryonic stem cell research has yet to yield any cures, what makes it momentous is that a new model of biomedical 'production' is being forged. As the Australian sociologist Cathy Waldby and her colleague Melinda Cooper have written, the stem cell industries rely heavily on the 'generative' potential of the female body as the source of eggs, placental cord blood and other tissue:

> *The stem cell sciences aim to transform this gene-rative capacity into regenerative capacity – to divert this productivity away from the generation of new individuals and toward the regeneration of existing populations.[15]*

A particularly controversial form of this way of profiting from female tissue is umbilical cord blood banking. Using what one commentator calls 'aggressive marketing techniques',[16] private firms charge prospective parents a substantial lump and/or yearly sum to store a portion of the mother's blood[17] that would normally flow to the baby through the placenta and umbilical cord (although typically the firms present the blood as a 'waste' product). The hope is to create a tissue-matched 'spare parts kit' on which the child can draw when (really 'if') stem cell technology is sufficiently advanced.

However, cord blood banking can pose risks to both mother and child if the cord is clamped too early, if too much blood is taken or if the attention of delivery room staff is diverted at a crucial moment of childbirth,[18] while the value to the baby is largely unproven except in comparatively rare instances. Two reports from the Royal College of Obstetricians and Gynaecologists have advised against the routine taking of cord blood, as has a survey of US physicians.[19] Likewise, a study published in October 2010 concluded that private cord blood banking only paid off when storage costs are less than £165[20] – about one-tenth of what some banks charge. Some countries continue to ban private banking for all these reasons, among them Italy, Belgium and (no surprises here) France.

Other private firms have sprung up more recently, offering a similar 'service' for other forms of tissue, including children's milk teeth. At least that doesn't involve any risk, except perhaps the risk of competition for the Tooth Fairy. Yet who can resist the myth of the infinitively regenerative body – especially when we hold such hopes not on our own behalf, but for our children? Such hopes are of course universal, and developing countries are beginning to follow Western trends in private cord blood banking, with entries into the sector from Latin America, India and China.

More generally, in other forms of stem cell research, the developing world is entering the race too. What the Asian expert Margaret Sleeboom-Faulkner calls 'a bioethical vacuum'[21] has allowed the stem cell industries in China to take advantage of George Bush's prohibition on federal funding and to develop a flourishing but ineffectively regulated sector. But she and her colleague Prasanna Kumar Patna warn that the populations in developing countries may become 'techno-coolies', in a new form of colonialism. That wouldn't be the first time biotechnology has used or abused Third World populations, as the next chapter will show.

7

Sacrificial lambs and professional guinea pigs: a bestiary of research ethics

It's 1946. In the dock on one side of the Atlantic, Nazi doctors are being tried at Nuremberg by US prosecutors for crimes against humanity, in the form of 'research experiments' carried out on concentration camp prisoners. And on the other side in Guatemala, the US Public Health Service (PHS) is deliberately infecting prisoners and mental patients with syphilis in another 'research experiment', with the goal of creating a prophylactic against the disease to replace the ineffective drugs used for soldiers during the war that just ended.

Sixty-three years later, an American historian, Professor Susan Reverby, was rummaging through archived medical papers from the 1940s. Reverby was completing a final task in her two decades of studying the PHS's detestable Tuskegee experiments – in which hundreds of African-American men with late-stage syphilis were observed but not treated, even after penicillin was developed. She was examining the papers of Thomas Parran, US surgeon-general from 1936 to 1948, when the Tuskegee research was already in full swing.

And so, she found, was the previously unknown Guatemalan 'experiment'. For years Tuskegee has been a byword for abuse of research ethics – to the extent that President Clinton apologized to its surviving 'subjects'. But Reverby was to find that, if such a thing were possible, Guatemala was an even more egregious abuse. As she said,

> *Flashing red lights. I'd spent nearly two decades explaining that there had been no inoculation at Tuskegee, that while the PHS had used deplorable*

> *ethics, they had never infected anyone with syphilis. And here it was . . . the US government's health service had deliberately infected 427 Guatemalan men and women, prisoners and mental patients, with syphilis.[1]*

The US prosecutors at Nuremberg didn't know about the Guatemala experiments at the time of the Nuremberg trials, so there's no allegation of deliberate hypocrisy. That's not the issue. Instead, the questions are:

▶ How were the public health authorities able to override the basic medical ethics rule of 'First do no harm'?
▶ Why didn't they think informed consent was necessary?
▶ How was a rich developed country able to persuade a weaker, poorer country to sacrifice its nationals – particularly when that involved very vulnerable populations, prisoners and mental patients, who have also been used as trial subjects in a number of other abuses of research ethics?[2]

Guatemala and Tuskegee were extreme examples of abysmal research ethics, and they happened many years ago (although the Tuskegee 'experiment' went on until 1972).[3] As the US bioethicist Art Caplan has said about the Guatemala case:

> *When you give somebody a disease – even by the standards of their time – you really cross the key ethical norm of the profession.[4]*

You might think that such an outrage couldn't happen any more. Bioethicists disagree among themselves about that, with Dan Brock at Harvard Medical School calling it 'pretty unlikely' but Eric Meslin at the University of Indiana warning that 'it could happen today'.[5] After President Obama, like his predecessor, issued an apology, he appointed a presidential bioethics commission to look at the adequacy of standards – which indicates that the president agreed with Meslin rather than Brock.

Tuskegee and Guatemala were examples of bad science as well as bad ethics, so it's not entirely fair to use them as examples of what untrammelled 'scientific curiosity' can lead to. But the issues about whether and how research ethics should be deployed to prevent science doing whatever it can do – that theme from Chapter 1 – keep on recurring, albeit in guises that make them harder to spot than blatant wrongs such as the Nazi atrocities, the Tuskegee case or the Guatemalan study. More typically, they now concern economic exploitation or cultural clashes rather than medical risk – although that element certainly hasn't disappeared, especially when poor and vulnerable subjects are tempted to accept risks for reward.

The Nuremberg Code

The Nuremberg trials resulted in the Nuremberg Code, which was supposed to prevent Nazi-style abuses from recurring by settling issues like informed consent and doing no harm, even though the code was voluntary. Its successor, the Declaration of Helsinki, has gone through several revisions but is still the agreed standard today. The fact that there are agreed standards puts research ethics ahead of other controversies in bioethics – such as genetic patenting, where the issues are still being dragged through the courts. Below is the crucial first paragraph of the Nuremberg Code:[6]

The voluntary consent of the human subject is absolutely essential. This means that the person involved should have the legal capacity to give consent, should be so situated as to be able to exercise free power of choice, without the intervention of any element of force, fraud . . . or other form of constraint or coercion; and should have sufficient knowledge and comprehension of the subject . . . to make an understanding and enlightened decision. This latter element requires that before the acceptance of an affirmative decision by the experimental subject, there should be made known to him the nature, duration and purpose of the experiment; the method and means by which it is to be conducted; all inconveniences and hazards reasonably to be expected and the effects on health which may possibly come from his participation in the experiment.

The duty and responsibility for ascertaining the quality of the consent rests upon each individual who initiates, directs or engages in the experiment.

Ironically, the Tuskegee and Guatemala experiments have actually been explained in terms of national solidarity. Along with studies such as the 1942 trial in which Michigan mental patients were deliberately exposed to flu after being injected with an experimental vaccine – a trial in which a younger Jonas Salk was co-investigator – they were ostensibly justified in terms of the urgency of conquering killer contagious diseases and benefiting the nation as a whole. In the words of Laura Stark, a US professor of science in society:

> *It was unusually unethical, even at the time . . . [But] there was definitely a sense – that we don't have today – that sacrifice for the nation was important.*[7]

We heard something similar at the time of Hwang Woo Suk's experiments, during the euphoria before the truth was revealed: that Korean women were willing, cheerful and patriotic sacrificial lambs. That's all very noble, but whose sacrifice is expected for whom? Why should poor African-American men have been expected to sacrifice their health, without even a by-your-leave – and for no predictable benefit? And why should Guatemalans make sacrifices for another nation altogether?

That contradiction is all the more acute now that a substantial proportion of clinical trials are 'outsourced' to the Third World – a phenomenon that outstrips 'reproductive tourism' and 'surrogacy contracting' to India, although the same dynamic is at work. A report in 2010 revealed that foreign citizens made up more than three-quarters of all the subjects in clinical trials conducted by US firms and researchers. The US Food and Drug Administration inspected only 45 of these sites, about 0.7 per cent. This isn't altogether a new phenomenon: it's been alleged that the PHS had deliberately moved the Guatemalan study abroad from Terre Haute, Indiana, because offshore testing wasn't subject to the same level of scrutiny.[8] The director of the Guatemalan research may have believed he was doing good science, but he also wrote to his own supervisor:

> *Well, all I can say is, it's against the law to do many things, but the law winks when a reputable man wants to do a scientific experiment . . . Unless the law winks occasionally, you have no progress in medicine.*[9]

There's no suggestion that Third World patients are deliberately being made ill when research is outsourced there these days – unlike in the Guatemalan case – but there have been controversies about whether populations lacking basic medical care are inherently vulnerable. Can they really make an informed choice about whether to enter a clinical trial when the choice may be either participating in the trial or receiving no medical treatment at all? The Nuremberg principles weren't and couldn't have been devised to deal with this sort of situation: the absence of genuine consent from the concentration camp prisoners was obviously of a different sort altogether.

During a major meningitis epidemic in 1996 in northern Nigeria, the drug company Pfizer supplied doctors with the oral antibiotic Trovan®, which the firm wanted to test against the most effective known drug, ceftriaxone, as a 'control'. This procedure is actually consistent with the Declaration of Helsinki and with the general consensus in research ethics that the control group has to receive the best known treatment for comparison.

The Trovan trial was arguably less controversial than an earlier case, in which pregnant African women with human immunodeficiency virus (HIV) were enrolled in trials about preventing transmission to their fetuses.

There, researchers were comparing a low-dosage anti-retroviral regime not with the best known treatment available in the West, but with a placebo – on the grounds that what was relevant was the best locally available treatment (nothing).

Even that trial provoked opposite reactions: proponents, including many African commentators, favoured the trial because if the cheaper low-dosage regime succeeded, it would be more likely to be taken up in Africa than the expensive 'gold standard' treatment.[10] But shouldn't the standards of care in clinical trials be universal, not dependent on the colour of the subjects' skin?

Yet the Trovan trial also stirred up a storm, for two different reasons. First, even if the trials were favourable, Trovan was never intended for sale in Africa, but rather in the USA and Europe. Second, the sparse clinical teams were already facing not one but three epidemics – measles and cholera as well – so both they and the children's families were desperate. In the words of Jean-Hervé Bradol, who was president of Médecins sans Frontières and in charge of the African teams:

> It was not a time for a drug trial at all. They were panicking in the hospital, overrun by cases on the verge of dying. The team were shocked that Pfizer continued the so-called scientific work in the middle of hell.[11]

Eleven children died during the Trovan trial, while others were permanently disabled; several families have brought a complaint, and Pfizer has settled with the Nigerian government (without admitting responsibility).

However, the death rate during the Trovan trial was well below the rate for untreated meningitis – and was much the same in both arms. Five children died on Trovan, six on the control drug (ceftriaxone).

So you could certainly argue that the children who were saved by the Trovan trial benefited, even if in future other African children stricken with meningitis weren't going to be offered the drug. And perhaps you could also speculate that the success of the trial, in those terms, made it more likely that there would be pressure on the drug company to lower its prices and make Trovan more widely available. That did happen with anti-retroviral drugs for HIV in South Africa.

But is that good enough? When trials are outsourced abroad, or delegated to private profit-making companies, how can the Nuremberg or Helsinki principles be monitored? In principle, regulation is supposed to be up to Institutional Review Boards (IRBs) at the principal investigators' host institutions (local research ethics committees in the UK) or to the grant-giving bodies behind them. Researchers often complain that regulation is, if anything, too constricting, stifling medical research and investment. In the words of a recent report from the UK's Academy of Medical Sciences:[12]

> [T]here is evidence that UK health research activities are being seriously undermined by an overly complex regulatory and governance environment . . . [A]fter funding for a study has been agreed, it now takes an average of 621 days to recruit the first patient. In short, the current situation is stifling research and driving medical science overseas.

This is a common complaint on both sides of the Atlantic, but not one that impresses the US bioethicist Carl Elliott. In his neatly titled book *White Coat, Black Hat: Adventures on the Dark Side of Medicine*, Elliott has explored the unregulated netherworld of professional guinea pigs. Like the African subjects of the Trovan trials, these Americans aren't going to reap any benefits from the drugs they test for a living, because they're poor and uninsured. That's why they're the only ones willing to become guinea pigs for phase I trials, which are designed to test toxicity.[13] But if those trials go wrong for them, only 16 per cent of US academic medical centres will cover their healthcare.[14]

Many IRBs, Elliott charges, are themselves also for-profit concerns, sometimes funded by the very companies they're meant to regulate. In one case, a 'sting' operation conducted by the General Accountability Office and a congressional committee discovered that a Colorado IRB was happy to approve a 'clinical trial' for a non-existent company wanting to test a fraudulent, unproven medical product.[15] In fact, over 70 per cent of US drug trials in 2004 were conducted in the regulation-lighter private sector.[16]

The dominance of private financial interests in publicly funded or targeted research is one thing that's changed since the Nuremberg Code. Another is the way in which informed consent is no longer just a matter of not experimenting on unwilling individuals. You'd think that might be simple, but it's not, particularly where cultures collide. We saw an example of that in the Tongan case in Chapter 4.

A more recent controversy has involved a biobank – a collection of stored tissue samples – taken from the Havasupai tribe in northern Arizona, who filed a $50 million lawsuit claiming that Arizona State University (ASU) researchers had improperly used their members' blood samples. They alleged lack of properly informed consent – along with fraud, breach of fiduciary duty, negligence and trespass.

Twenty years earlier, ASU scientists had collected over 200 blood samples from tribal members, in what the tribal elders had been told was a study of diabetes. The consent form was broader – too broad, in fact, with its vague wording about studying 'the causes of behavioral/medical disorders'. And true to the waffly wording of the consent form, the researchers used the samples for multiple studies unrelated to the research on diabetes – which hadn't panned out – without informing the Havasupai.

Tribal members later discovered – and objected to – use of the samples in studies on schizophrenia and inbreeding. They claimed that the resulting two dozen published scientific papers had damaged and stigmatized the tribe.[17] While this objection was widely reported in the press as having to do with Havasupai cultural beliefs, anyone might be nervous about being stigmatized as schizophrenic or inbred in published articles with the apparent authority of science behind them. But the Havasupai did also object to evolutionary genetics studies performed on the samples, throwing the tribe's origin stories into doubt.

The Havasupai couldn't really have expected to win: a number of precedents were against them. Other groups or individuals have also challenged 'blanket' consents to use of their tissue or data in biobanks for other purposes than they intended, and they've generally lost.[18] Nor, as we saw earlier, does the common law recognize that you can have property rights in tissue once it's left your body.

But the university did (eventually) apologize, settle out of court for $700,000 and agree to return the remaining samples to the tribe. This sets a possible legal precedent (although there was no court judgment) and does raise difficulties for researchers about whether they need to obtain a fresh informed consent for every new use of stored samples. But then the question is whether the initial 'consent' was really 'informed'.

Researchers might be forgiven for thinking that the principles of research ethics only require them to be open and honest about medical risks. That's what the Nuremberg Code is concerned with:

> 'Informed consent' requires that before the acceptance of an affirmative decision by the experimental subject, there should be made known to him the nature, duration and purpose of the experiment; the method and means by which it is to be conducted; all inconveniences and hazards reasonably to be expected and the effects on health which may possibly come from his participation in the experiment.

That's actually quite a lot to do, without worrying about social or cultural risks such as stigmatization. On the other hand, the intent of the Nuremberg Code is clearly to make researchers be as open and specific as possible with their subjects – and a vague blanket consent form ('the causes of behavioral/medical disorders') certainly goes against that grain.

The dilemmas of biomedical research haven't lessened since Nuremberg: like pluripotent stem cells, they've just changed shape. What has changed is the public climate in which research takes place. Jean McHale, a British professor of medical law, thinks our attitudes have swung round 180 degrees since Nuremberg: now, she says, we assume research is never wrong.[19] The claim that we have a positive duty to participate in medical research demonstrates exactly that attitude.

You could almost say we have a quasi-religious attitude to medical research. And that leads us nicely into the concerns of the final chapter: 'God, Mammon and biotechnology.'

8

God, Mammon and biotechnology

The notion that science and religion are at war is one of the great dogmas of the present age.[1]

ALL THAT MATTERS

It is a dogma universally acknowledged, as well – to paraphrase the opening lines of *Pride and Prejudice*. We first met it during the 'cybrids' clash described in Chapter 1. But while this belief might just about be excusable in those who deal in dogma – religious fundamentalists – it should have no place whatsoever in the belief system of committed rationalists, in my view. Yet it often seems as if those who promulgate this dogma most assiduously are the scientific rationalists.

In *The God Delusion*, Richard Dawkins asks, 'What's wrong with religion? Why be so hostile?'[2] He goes on to answer his own question: 'As a scientist, I am hostile to fundamentalist religion because it actively debauches the scientific enterprise.'[3] Although he says he has a 'dislike of gladiatorial contests',[4] he feels he has little option but to defend science against religion in what he apparently accepts is something very much akin to a fight to the death.

I think this is a shame and an error. What 'actively debauches the scientific enterprise', in my view, is not religion, but the commercialization of biomedicine: not God, but Mammon.

It's not Opus Dei that controls over 70 per cent of US drug trials: it's for-profit companies. It's not the Vatican that holds research-blocking patents on human genes: it's for-profit companies. It wasn't Bible-bashers who launched a four-year libel action against Dr. Peter Wilmshurst, after he gave a scientific conference paper questioning the usefulness of a particular cardiological device. It was the company that made the device.

While it's true that US stem cell research was temporarily impeded by George Bush's prohibition on federal funding for embryonic stem cell lines, scientists found ways around that ban: the creation of the multi-million-dollar California Institute of Regenerative Medicine, the import of foreign stem cell lines and the development of alternative, non-embryonic forms of stem cells. It's not all that clear, anyway, that Bush's opposition to embryonic stem cell research was motivated by his born-again Christianity. Melinda Cooper, a sociologist of science, thinks that Bush pulled off a political tour de force by mollifying the evangelical right while also making Big Biotech very happy – particularly the corporation that held all the most important currently available stem cell lines, whose value shot up as a result of the ban.[5]

Either way, scientists were still able to make progress in stem cell research, despite what has been billed as the biggest victory of religion over science in recent times. In fact, it's been said, rightly or wrongly, that without the Bush ban there wouldn't have been any incentive to develop induced pluripotent stem cell lines, which now look like a major resource. But no scientist can invent around a gene patent by a private firm determined to block access to any rival who might develop a cheaper drug. If you can't get at the gene, you can't do the research, full stop. Mammon 1, scientists 0.

So who's out there defending the fort? In the Myriad case (Chapter 5), medical professional organizations made common cause with civil liberties groups and individual patients against restrictive genetic patents. But some of the most valiant defenders against abuses of biomedicine haven't been scientists – many of whom are of course funded directly or indirectly by business – or even bioethicists, although many senior bioethicists were attracted to that new field as an extension of their 1960s activism.[6] No: it's been the tiny, not-for-profit, activist organizations like Korean Womenlink, which spilled the beans on Hwang Woo Suk, or GeneWatch, which succeeded in driving a 'nutrigenomic' product off the UK market by demonstrating that the manufacturer's claims were unfounded.[7]

But very few people know about them, whereas many know about *The God Delusion*. So the dogma about religion and science being at war is a red herring, straw man, blind alley or whatever trite metaphor you prefer. It has had the (doubtlessly unintended) effect of diverting critical inquiry in biotechnology from the most important bits of evidence – ironically enough, since evidence and truth are sacred to its acolytes.

In this book, I've tried to rectify that balance as best I can. So in this chapter, I'm not primarily concerned with the question that has been tackled by many other writers – whether science is compatible with religion. That, too, is a diversion from the economic and political analysis of whose interests are served by maintaining the dogma that science and religion will never be reconciled. I'm more interested in biotechnology's relationship with Mammon than with religion, although I certainly hold no brief for Opus Dei, the Vatican or Bible-bashers. But before proceeding with that analysis of whose interests are at stake, it's worth a quick look at two main lines of argument against the view that science and religion are at war.

First, many prominent scientists have asserted that there is no necessary opposition between the scientific enterprise and the religious mindset. In 1941, Einstein famously declared:

> Science can only be created by those who are thoroughly imbued with the aspiration toward truth and understanding. This source of feeling, however, springs from the sphere of religion . . . Science without religion is lame, religion without science is blind.[8]

Galileo's mistreatment by the Church has given rise to the misapprehension that he attacked faith itself, but actually he believed that truth was accessible by both scientific inquiry and religion. This view is surprisingly similar to that of a more modern scientist, Stephen Jay Gould, who has described religion and science as two separate and non-overlapping 'magisteria'. Because they're completely divorced from each other, Gould thinks, they don't threaten each other:

> The net, or magisterium, of science covers the empirical realm: what is the universe made of (fact) and why does it work this way (theory). The magisterium of religion extends over questions of ultimate meaning and moral value. These two magisteria do not overlap.[9]

Dawkins has criticized Gould's view as 'dishonest', because it fails to recognize that religion does make truth-claim statements about the natural world. I think that it's wrong for the opposite reason: it implies that bioethics, which, like religion, concerns questions of moral value, can have nothing to say about science. But science also makes implicit truth-claims about ethics, including that by-now familiar claim that science should be allowed to proceed unregulated, so that we will all benefit. Besides being a philosophical position (a utilitarian one) rather than a scientific fact, this is a prescription for letting science do whatever it can do, and I have already dismissed that view in the first chapter.

More recently, prominent scientists such as Francis Collins have 'come out of the closet' about their own religious beliefs, arguing the same position from a personal perspective. Professor Jocelyn Bell Burnell, who should have won the Nobel prize for discovering pulsars (it went to her male supervisor instead), recently gave a public lecture in Oxford in which she likewise discussed whether and how her Quakerism is compatible with her scientific beliefs.[10]

This, then, is the second line of argument against the idea that science and religion are inevitably opposed. While some scientists (such as Francis Crick of double helix fame) have thought that 'the facts of science are going to make us become less Christian',[11] other equally eminent researchers have found space for religious belief in their personal lives. Social scientific surveys have corroborated these individual accounts, including a recent survey of 1,600 US scientists – only 15 per cent of whom thought there was an irreconcilable conflict between science and religion.[12] But the problem with this approach is that it descends into an unproductive balancing exercise. How many Collinses and Burnells does it take to trump a Dawkins and a Crick?

The founder of modern empirical inquiry, Francis Bacon, picked the original fight with representatives of religion, but also thought that there was no inherent incompatibility between the pursuit of knowledge and the pursuit of holiness. Although he pilloried 'the ignorance and error', 'the zeal and jealousy of divines', he also wrote:

> [O]ur Saviour himself did first show His power to subdue ignorance, by His conference with the priests and doctors of the law, before He showed His Power to subdue nature by His miracles. And the coming of this Holy Spirit was chiefly figured and expressed in the similitude and gift of tongues, which are but vehicula scientiae (vehicles of science or learning).[13]

Likewise, he wrote that 'the highest link of Nature's chain must needs be tied to the foot of Jupiter's chair'.[14] But while Bacon denied an irrevocable divide between science and religion, he did warn against what he prophetically saw as the real bane of the new science – worshipping Mammon:

> For men have entered into desire of learning and knowledge, sometimes upon a natural curiosity and inquisitive appetite; sometimes to entertain their minds with variety and delight; sometimes for ornament and reputation; and sometimes to enable them to victory of wit and contradiction; and most

times for lucre and profession, and seldom sincerely to give a true account of their gift of reason to the benefit and use of men, as if there were sought in knowledge a couch whereupon to rest a searching and restless spirit; or a terrace for a wandering and variable mind to walk up and down with a fair prospect; or a tower of state for a proud mind to raise itself upon; or a fort or commanding group for strife and contention; or a shop for profit or sale; and not a rich storehouse for the glory of the Creator and the relief of man's estate . . . I do not mean, when I speak of use and action, that end before mentioned of the applying of knowledge to lucre and profession; for I am not ignorant how much that diverteth and interrupteth the prosecution and advancement of knowledge.[15]

In this book, we have seen a number of ways in which 'the applying of knowledge to lucre' and the commodification of biotechnology have 'diverted and interrupted the prosecution and advancement of knowledge' – such as the restrictive patenting cases.

More generally, research agendas are increasingly dictated by what drugs it will be profitable to produce, rather than where the most benefit can be obtained (the issue in the Trovan case). GeneWatch also highlighted this issue in its campaign against the nutrigenomic manufacturer Sciona, alleging that the firm's main concern was which genes were likely to yield marketable products for the worried Western well, not which diseases were causing the highest mortality. The late US sociologist of science Dorothy Nelkin warned just before her death in 2003 that biomedical research, particularly in genetics, was increasingly dictated by corporate agendas.[16] If that continues, the scientific method will itself become nothing more than a commodity on sale to the highest bidder.

So what has bioethics done about the risk that biomedical science may degenerate into 'a shop for profit or sale'? Not enough, many of us in the field think (though not as many as of us as should think so). The American medical sociologist John Evans argues that the field of bioethics is no longer critical and independent, but rather 'inside the belly of the whale' of commercial modern biotechnology.[17] Theology and religion have been discredited as counterweights to neo-liberal arguments in favour of deregulated biotechnology, he thinks, but bioethics hasn't developed its own firm identity to replace it. As he writes:

> *Of course the self-image of the bioethics profession is that it is an opposition movement to the power of scientists and physicians, and it did truly force its ethical system on scientists and physicians in the 1960s and 1970s. But I concur with historian Charles Rosenberg that 'as a condition of its acceptance, bioethics has taken up residence in the belly of the medical whale; although thinking of itself as still autonomous, the bioethical enterprise has developed a complex and symbiotic relationship with this host organism. Bioethics is no longer (if it ever was) a free-floating, oppositional and socially critical reform movement'.[18]*

There is a wince-inducing truth in this charge. But Rosenberg wrote that in 1999, and both biotechnology and biomedicine have moved on since then, as has society at large. There is much more concern about commodification and corruption in biomedicine,[19] while at the same time the neo-liberal pro-privatization consensus and state spending cuts have been contested everywhere from Syntagma Square in Athens to the Occupy camps in London and New York. Critiques of privatization and corporate dominance of the state, such as Naomi Klein's *No Logo* and *The Shock Doctrine*,[20] have become bestsellers, while plays such as *Enron* have unexpectedly wound up as Broadway hits. In general, we have a lot less reason to be naïve – although sometimes we still fall for scientific 'miracles' like Hwang's 'research'.

We – bioethicists and the public alike – just need to grow up. That maturation will come from putting away childish things – like the bogeyman about religion being the biggest enemy of scientific progress, or the puerile fantasies of 'enhancement' – and recognizing what we need to do to make biomedicine into Bacon's 'rich storehouse' for everyone. There are already movements in that direction, for example the Tarrytown meetings of 2010–12, 'an annual convening of scholars, policy experts, advocates and others working to ensure that powerful new human biotechnologies and related emerging technologies support rather than undermine social justice, equality, human rights, ecological integrity and the common good'.[21]

It's not going to be simple. How could it be, with the forces of 'biocapitalism' arrayed against us? But bioethics, properly conceived and executed, unites within itself a rare and powerful combination of expertises in philosophical analysis, legal and political acumen and scientific awareness. We're in a strong position to help rather than hinder the best that science can propose. We – and I count you, the reader, as one of us – are in a strong position to help science do its best for the common good.

IDEAS 100

20 suggestions for further reading

1 George Annas, *Worst Case Bioethics: Death, Disaster and Public Health* (Oxford University Press, 2010). Annas is one of the few English-speaking bioethicists who has consistently put social justice above individual autonomy. His latest book compares the over-optimism of the enhancement debate, which he sees as assiduously promoted by business interests in biomedicine, with the neglect of real priorities such as pandemics and public health.

2 James Boyle, *Shamans, Software and Spleens: Law and the Construction of the Information Society* (Harvard University Press, 1996). Boyle's influential analysis likens the patenting of large portions of the human genome to the agricultural enclosures of common land before the

Industrial Revolution. In both, things previously outside the market were colonized and taken over for private profit.

3 **Howard Brody,** *The Future of Bioethics* (Oxford University Press, 2009). Like Sherlock Holmes's curious incident of the dog in the night-time, the clue to bioethics, Brody thinks, is why its dogs don't bark when they should. He accuses mainstream bioethicists of being too wary of speaking truth to power and of failing to consider justice rather than autonomy as the primary value.

4 **Allen Buchanan,** *Beyond Humanity?* (Oxford University Press, 2011). Buchanan, a moderate advocate of enhancement, presents a carefully selected and well-balanced range of what he calls 'anti-anti-enhancement' arguments.

5 **Melinda Cooper,** *Life as Surplus: Biotechnology and Capitalism in the Neo-Liberal Era* (University of Washington Press, 2007). This brilliant book by a young scholar with a very wide-ranging mind 'connects the utopian polemic of free-market capitalism with growing internal contradictions of the commercialized life sciences', as the back cover summary puts it.

6 **Donna Dickenson,** *Body Shopping: Converting Body Parts to Profit* (Oxford: Oneworld, 2009). Along with a survey of the commercialization of human tissue from BC to AD – before conception to after death – Dickenson offers explanations of legal cases, narratives of personal patient accounts, and an original analysis of the way in which all bodies are 'feminized' by being treated as objects in the free market in tissue.

7 **Donna Dickenson, Richard Huxtable and Michael Parker,** *The Cambridge Medical Ethics Workbook,* 2nd edition (Cambridge University Press, 2010). In this unusual and accessible textbook, which includes a CD-ROM with six interactive filmed scenarios, the authors use a 'bottom-up' approach to introduce medical ethics through real-life, everyday cases.

8   Carl Elliott, *White Coat, Black Hat: Adventures on the Dark Side of Medicine* (Beacon Press, 2010). Elliott, himself a medically trained bioethicist, draws a telling comparison between the internet and medicine: both are highly commercialized, but everyone knows that's true of the net, whereas people still trust medicine. He shows how medical research has been increasingly outsourced to the Third World or to private companies, in a system where both scientists and subjects are motivated by profit, not altruism.

9   Cecile Fabre, *Whose Body Is It Anyway? Justice and the Integrity of the Person* (Clarendon Press, 2006). Distributive justice demands that the sick should have the right to be given or even to buy the organs of those who are healthy, Fabre asserts, just as the poor have rights to share in the wealth of the rich – but while her motivation is fairness, she wrongly assumes that we own our bodies as we do other forms of wealth.

10 Kieran Healy, *Last Best Gifts: Altruism and the Market for Human Organs* (University of Chicago Press, 2006). A well-documented sociological analysis of the complexity of distinctions between altruism and markets, giving and selling, in relation to organ transplants.

11 Jonathan Herring, *Medical Law and Ethics*, 3rd edition (Oxford University Press, 2010). A lively and comprehensive introduction to medical law which includes an online resource to enable the reader to keep up to date with changing cases.

12 Sheila Jasanoff, *Designs on Nature* (Harvard University Press, 2005). In what the book jacket calls a 'magisterial look at some twenty-five years of scientific and social development', Jasanoff compares the politics of the life sciences in Britain, Germany and the USA. Like most US scholars she suffers from starry-eyed syndrome about the quality of regulation in Britain, but this is still a useful background book on policy-making in bioethics during its formative period.

13 Mary Midgley, *Science as Salvation* (Routledge, 1992). Whether we should let science do whatever it can do is bound up with the question of whether belief in the benefits of science is the modern substitute for Christian belief in redemption. As Midgley writes, 'The idea that we can reach salvation through science . . . is by no means nonsense, but it lies at present in a good deal of confusion.'

14 Dorothy Nelkin and M Susan Lindee, *The DNA Mystique: The Gene as Cultural Icon* (W W Norton, 1995). In this classic book, Nelkin and Lindee assert that popular and media understandings of the new genetics treat DNA as the equivalent of the human soul: the sacred essence of the person and of human existence.

15 Michael Sandel, *The Case against Perfection: Ethics in the Age of Genetic Engineering* (Harvard University Press, 2007). In this extended version of an influential essay that originally appeared in the *Atlantic Monthly*, the political theorist Michael Sandel argues that infatuation with 'enhancement' weakens our sense of solidarity, responsibility and humility.

16 Debra Satz, *Why Some Things Should Not Be for Sale: The Moral Limits of Markets* (Oxford University Press, 2010). Why should a surrogate mother have to hand over the baby, when no other kind of employment contract can force the worker to actually perform the job and not just return the money he was paid? Satz perceptively picks up this crucial point about 'specific performance' in her careful economic and legal analysis.

17 Rebecca Skloot, *The Immortal Life of Henrietta Lacks* (Crown Books/Pan Macmillan, 2010). With considerable patience and sensitivity, Skloot traces the fatal consequences for the Lacks family of the infamous 1950s case in which immortal cell lines, still used widely in scientific research, were created with cancer tissue taken without consent from an impoverished African-American woman before her death.

18 Richard Titmuss, *The Gift Relationship* (Allen and Unwin, 1970, reprinted LSE Books, 1997). Although markets in blood and other forms of human tissue have changed radically since Titmuss wrote this classic book, and although his concept of the gift is out of academic vogue, this is still the crucial analysis from which to begin.

19 Harriet A Washington, *Deadly Monopolies: The Shocking Corporate Takeover of Life Itself – and the Consequences for Your Health and Our Medical Future* (Doubleday, 2011). An up-to-date exposé of corporate commodification in genetic patenting and other areas of biomedicine.

20 Catherine Waldby and Robert Mitchell, *Tissue Economies: Blood, Organs and Cell Lines in Late Capitalism* (Duke University Press, 2006). Waldby and Mitchell demonstrate just how shaky are some of the structures underpinning the global politics of human tissue, such the distinction between gift and commodity, when even gift is commodified on a worldwide scale – leaving the human body as an open source of free tissue for commercial use.

10 landmark court decisions

21 Diamond *v* Chakrabarty (447 US 303, 1980). The first case to establish that a patent can be granted on a living organism. In the case of a genetically engineered bacterium used to break down crude oil, the court ruled that 'Anything under the sun made by man' is patentable, with emphasis on the 'made' – even though the inventor himself said he had simply shuffled the genes and changed something that already existed. Now one in five human genes is the subject of a patent, mostly by private companies.

22 In the matter of Baby M (270 NJ Supr 303, 1987). The court set a precedent on which the 'surrogacy' industry built, by ruling in favour of the genetic father in a case where the birth mother (who was also the genetic mother) decided

she wanted to retain custody of the child. The precedent was ambiguous, however: the court also ruled the contract invalid – on the patriarchal grounds that the genetic father 'cannot contract for what is already his' – and awarded the father custody on the grounds of supposed 'best interest of the child', while giving the mother visitation rights.

23 **Moore _v_ Regents of the University of California** (51 Cal 3rd120, 793 P 2d, 271 Cal Rptr 146, cert. denied 111 SCt 1388, 1990). The landmark case about property in the body. Moore failed to establish that he had a property right in tissue taken from his spleen and other sites, which was then developed into a valuable cell line, although the court agreed that he had not given a properly informed consent. (Indeed, the court agreed that he was lied to when he was told that the ongoing tissue removals were vital for his leukaemia treatment.) The majority feared that allowing Moore a property right would contravene the traditional common law doctrine that excised tissue is _res nullius_ ('no one's thing'), along with impeding scientific research and creating a marketplace in body parts. In his dissent, however, Justice Broussard remarked that there was already a marketplace, one in which everyone stood to make a profit from Moore's tissue except Moore.

24 **Howard Florey/Relaxin** (European Patent Office Reports 541, 1994). Relaxin is a protein secreted by pregnant women which eases childbirth. The morality of the patent by Howard Florey was questioned in 1994, on the basis that the removal of tissue from pregnant women was an affront to human dignity because it used pregnancy for profit. But the Opposition Division of the European Patent Office ruled that the tissue had been taken with women's consent and that individual consent rather than commercial profit was the sole issue.

25 **R _v_ Kelly** (3 All ER 741, 1998). Kelly was a sculptor who colluded with a colleague to take preserved body parts

from the Royal College of Surgeons without the doctors' consent. In his defence he made the cunning and brazen claim that he had not committed theft because there is no such thing in common law as property in the body. The College succeeded in recovering the body parts by using the Lockean argument that their members had invested skill and labour in them.

26 **Greenberg** *et al. v* **Miami Children's Hospital Research Institute** (208 F Supp 2d 918, 2002). The Greenbergs had raised funds, contributed their dead children's tissue and enrolled other families in a research project on the fatal genetic condition of Canavan disease. Unbeknownst to them and the other families, however, the research institute took out a comprehensive patent on the gene coding for the disease and began to collect royalties from the patented test, as well as to restrict the number of laboratories licensed to perform the diagnostics. The Greenbergs and other families alleged that they would have withheld their crucial contributions if they had known that the hospital intended to commercialize the discovery and to restrict access for parents who could not pay the costs of the test, but they failed in their attempt to allege breach of informed consent, breach of fiduciary duty and fraud by the researchers. They did succeed in a claim about unjust enrichment, however, so this case is not as clear a victory for the forces of commercial biomedicine as Moore *v* Regents of the University of California.

27 **Washington University** *v* **Catalona** (437 F Supp 2d, ESCD Ed Mo, 2006). Catalona was a consultant and researcher who had developed a leading diagnostic test for prostate cancer. When he changed universities, he wanted to take his archive of prostate tissue samples with him for further research, and obtained his patients' consent to do so. His original university, however, thought the biobank was too commercially valuable to lose. They successfully brought an action establishing that the archive belonged

to them and that the patients had no ownership interest that would allow them to direct how their tissue should be used. A huge number of research institutions filed briefs in support of the winning university, indicating that major commercial interests were at stake.

28 **Yearworth v North Bristol NHS Trust** (3 WLR 1218, 107 BMLR 47, 2009). This case made legal history when the court found in favour of a property claim from several men whose sperm samples, taken before their treatment for cancer, had been negligently lost or destroyed by the hospital. 'Developments in medical science', the judges held, 'now require a re-analysis of the common law's treatment of and approach to the issue of ownership of parts or products of a living human body'. However, there were no commercial interests involved here.

29 **Tilousi v Arizona State University** (filed 2005, case settled out of court 2010). The Havasupai Indians of Arizona have an exceptionally high rate of diabetes – 55 percent in women and 38 per cent in men – so they readily co-operated with an Arizona State University researcher who wanted to take tissue samples from them to establish whether there was a genetic basis for their vulnerability. Without the participants' knowledge, however, the researcher also studied their samples for schizophrenia, inbreeding and ancestry. She and her colleagues published 15 articles on their findings, which the tribe regarded as degrading and offensive. They went to court to establish that the research was unauthorized and that they had control rights over the use of their samples. Although the court found that they had given broad consent to any purpose of the research, the university agreed to settle out of court.

30 **Association for Molecular Genetics v US Patent and Trade Office** (No. 2010-1406, Fed. Cir., 29 July 2011). This hotly fought case turned on the legitimacy of the *BRCA1* and *BRCA2* gene patents held by the firm Myriad Genetics, which charged a fee of over $3000 to women who wanted to have the diagnostic test for the variants of these genes that are implicated in some breast cancers. Because Myriad held a monopoly patent on the genes themselves, no other researchers could develop a cheaper test. In 2010 a lower court in New York unexpectedly agreed with the contention by a coalition of medical, patient and civil liberties groups that these gene patents were not allowable. 'Pigs returned to earth', as an industry newsletter put it, in July 2011, when an appeal court held by a 2:1 majority that the patents should be allowed. In March 2012 the US Supreme Court sent the case back to the appeal court for reconsideration in light of a subsequent case, but no one knows whether the outcome will change.

10 literary works

31 Lori Andrews, *Sequence* (2006). Lori Andrews is a professor of medical law who has been involved with almost every legal case about the limits of commercialization in biotechnology – most recently the Myriad Genetics case about whether patents on human genes are valid (see Landmark Court Decisions no. 30). As if that weren't enough, she's also written several thrillers about genetics and biotechnology, of which this is the first.

32 Margaret Atwood, *Oryx and Crake* (2003). Snowman is the last human survivor in a world populated by the 'Crakers', created by the genetic engineer Crake to serve humanity in the 'Paradise Project'. But, following 'the disease', nothing of that remains. Snowman, himself once a scientist, begins a dangerous journey to 'the Bubble', the epicentre of the enhancement project, to find out how disaster unfolded. The novel's plot may sound hackneyed, but the hellish atmosphere is masterful.

33 Matthew Cobb, *The Egg and Sperm Race: The Seventeenth-Century Scientists Who Unravelled the Secrets of Sex, Life and Growth* (2006). A lively and often surprising history of how far science has travelled in its understanding of reproductive biology from the days when people believed in the homunculus (a miniature person supposedly fully formed inside the human egg).

34 Sarah Hall, *The Carhullan Army* (2007). Reversing Margaret Atwood's *The Handmaid's Tale* (see Films no. 42), in this future dystopia set in a decaying Lake District town, women are compulsorily fitted with contraceptive devices. Yet neither Hall nor Atwood foresees the way in which women's reproductive capacity is actually being governed in our time – not by a dictatorial state but by commercial interests such as egg 'donation' agencies or 'surrogacy' brokers.

35 Nathaniel Hawthorne, 'Rappacini's Daughter' (1844). In medieval Padua Doctor Giacomo Rappacini uses his scientific knowledge to give his daughter Beatrice 'marvellous gifts against which no power nor strength could avail an enemy', but in 'the fatality that attends all such efforts of perverted wisdom', succeeds only in making her poisonous to anyone on whom she so much as breathes. The story ends with Beatrice's death from an antidote, Rappacini's rival and neighbour, Professor Pietro Baglioni, 'looked forth from his window, and called loudly, in a tone of triumph mixed with horror, to the thunderstruck man of science, – "Rappacini! Rappacini! And is this the upshot of your experiment?" '

36 Annabel Lyon, *The Golden Mean* (2010). If modern scientific inquiry's father was Francis Bacon, its grandfather is usually thought to be Aristotle. In this understated historical novel, cleanly written in the first person, Aristotle is depicted as more of a fumbler than a wise man, and yet as a determined scientist too.

37 Christopher Marlowe, *The Tragedy of Doctor Faustus* (1588). Although Faustus is prey to the usual temptations of the flesh – such as his wish to see the beautiful Helen of Troy restored to life – his real sin, in Elizabethan thinking, was his intellectual arrogance in wanting infinite scientific knowledge. In the same era that produced Francis Bacon and the origins of empirical science, this cautionary tale warns: 'Regard his hellish fall/Whose fiendful fortune may exhort the wise/Only to wonder at unlawful things/Whose deepness doth entice such forward wits/To practice more than heavenly power permits.'

38 David Mitchell, 'An orison of Sonmi-451', in *Cloud Atlas* (Sceptre, 2005). Technological dystopias may be a dime a dozen, but this is no ordinary one: it is far more unpredictable and chilling, not least because it is embedded in a book that contains more standard, realistic narratives. In her last moments before execution, the fabricant Sonmi-451 describes to the impersonal Archivist her life with members of her stem-line as a server who 'knew no mother or father but Papa Song, our corp Logoman'.

39 Marge Piercy, *Body of Glass* (1991). Alternating between a future in which it's not safe to swim in the sea because you might be harvested for your organs, and the Jewish ghetto in Prague in 1600, this ambitious novel uses the legend of the dybbuk, a mythical cyborg, to explore fantasies about genetic engineering and enhancement.

40 Richard Powers, *Generosity* (2009). Researcher and enhancement advocate Thomas Kurton, who claims that we can discover the genetic basis of happiness, tests his theory on a young woman, Thassa, whose serenity is untroubled by the traumatic experiences she underwent during the Algerian civil war. Convinced that ageing is not just a natural phenomenon but 'the mother of all maladies', Kurton comes across as the good guy beset by troublesome quibbling bioethicists.

10 films

41 Never Let Me Go (2010), from the novel by Kazuo Ishiguro: a twist on the coming-of-age movie, told from the viewpoint of a young woman who has been raised as part of a class of human clones to be sacrificed or 'completed' for their organs.

42 The Handmaid's Tale (1990), from the novel by Margaret Atwood: a dystopian look at 'surrogacy' in which their reproductive capacity is wrung from a subordinate class of 'handmaids' on pain of death.

43 Mary Shelley's Frankenstein (1994), a remake starring Kenneth Branagh as the scientist obsessed with taking science to its limits – creating life, if you can call it that.

44 Fixed (2011), a documentary by Regan Brashear that examines both disability and enhancement technologies.

45 Dirty Pretty Things (2002). The clinical details are completely implausible, but this fictional exposé of an illicit organ trade and the grim world of a Nigerian hotel maid does pack a punch.

46 Wall-E (2008). After its inhabitants' over-zealous love affair with technology has left the earth a polluted slum, one of the few surviving robots left behind to clear up the mess becomes a hero in this animated film. A big issue in the enhancement debate is whether, if humans ever became more like androids, they would lose their moral compass. In *Wall-E* the surviving humans have all the moral compass of an overfed slug, whereas the robot is responsible, affectionate and dutiful.

47 The Invisible Man (1933). Both horror and comedy, this film, like Frankenstein, is either one of the classics about what happens when we do whatever science lets us do, or unscientific tosh – possibly both.

48 **Made in India** (2011). This documentary by Rebecca Haimowitz and Vaishali Sinha follows a Texas couple who believe their Indian 'surrogate' is receiving a fee of $7,000 to carry twins for them, in addition to the fee they have paid to the broker. We learn that she is actually only seeing $2,000 of the money, but, shabby though that is, the real shock of the film comes from the viewer's dawning realization of the obvious risk that small-statured Indian women are undergoing in bearing babies fathered by big-boned Westerners – especially twins. The surrogate will have a Caesarean if necessary, paid for by the clinic – but what about the increased risk she'll be running when she has her next non-paid-for baby at home without medical supervision?

49 **Inherit the Wind** (1960). Based on the Scopes 'monkey trial', in which a Tennessee schoolteacher was prosecuted for teaching Darwinian evolutionary theory, this drama pits the forces of scientific progress against prejudiced fundamentalists. (The film opens with the music 'Gimme that old-time religion'.) Progress loses.

50 **The Man in the White Suit** (1951). In this Ealing comedy, Alec Guinness plays a naïve inventor who assumes that his creation, a fabric that never wears out, will be welcomed as a boon to society. Instead it makes him enemies everywhere – among garment workers threatened with unemployment and bosses threatened with bankruptcy. Their answer to 'should we do whatever science lets us do?' is a vociferous 'no'.

10 best websites

51 Biopolitical Times, www.biopoliticaltimes.org

52 Bioedge, www.bioedge.org

53 Bionews, www.bionews.org.uk

54 Hastings Center 'Bioethics Forum', www.thehastingscenter.org/bioethicsforum

55 Sense about Science, www.senseaboutscience.org

56 RH Reality Check, www.rhrealitycheck.org

57 Philosophy Bites www.philosophybites.com, also available through iTunes

58 Genethique (for those who read French),www.genethique.org

59 The New Atlantis, www.thenewatlantis.com

60 Little Atoms radio programme, www.littleatoms.com

10 thinktanks and activist organizations

61 The Center for Genetics and Society, 'a non-profit information and public affairs organization working to encourage responsible uses and effective societal governance of the new human genetic and reproductive technologies . . . The Center works in a context of support for the equitable provision of health technologies domestically and internationally; for women's health and reproductive rights; for the protection of our children; for the rights of the disabled; and for precaution in the use of technologies that could alter the fundamental processes of the natural world' (from their website, www.geneticsandsociety.org). They also publish the very useful weekly news round-up and analysis, *Biopolitical Times.*

62 The Hastings Center, 'a non-partisan research institution dedicated to bioethics and the public interest since 1969' (from their website, www.thehastingscenter.org). They also publish a very well-established journal, *Hastings Center Report*, as well as a *Bioethics Briefing Book.*

63 The Nuffield Council on Bioethics, 'an independent body that examines and reports on ethical issues in biology and medicine. It was established by the Trustees of the Nuffield Foundation in 1991, and since 1994 it has been funded jointly by the Foundation, the Wellcome Trust and the Medical Research Council' (from their website, www.nuffieldbioethics.org). The Council publishes period reports on topical issues, following consultations by groups of expert advisors.

64 The Corner House (www.thecornerhouse.org.uk), an independent campaigning organization that punches well above its weight on issues of the environment, corruption and bioethics.

65 Our Bodies Ourselves (www.ourbodiesourselves.org), founded in 1970 (originally as the Boston Women's Health Book Collective) to act as an advocacy and information centre on women's reproductive rights.

66 GeneWatch (www.genewatch.org), a non-profit activist organization that provides impeccably researched briefing documents, often containing information not generally available to the public, obtained through use of the Freedom of Information Act or litigation.

67 The Alliance for Humane Biotechnology (www.humanebiotech.com), a Californian activist organization working in the areas of reproductive and genetic technologies, human egg harvesting, cloning research, disability rights, biotechnology patenting, human–animal hybrid research and synthetic biology. Recently the Alliance has made considerable progress in getting agreement from the in vitro fertilization industry to set up the first national registry of egg donors, so that the long-term risks of ovarian hyperstimulation can be monitored.

68 The Council for Responsible Genetics (www.councilfor-responsiblegenetics.org), a coalition of scientists, public health activists and reproductive rights advocates, which led the successful effort to establish the US Genetic Information Nondiscrimination Act.

69 Singularity University (singularityu.org), co-founded in 2008 by Google, Nokia, Cisco and other technology corporations, along with NASA, 'to assemble, educate and inspire leaders who understand and develop exponentially advancing technologies to address humanity's Grand Challenges' [original capitals]. It pursues the transhumanist/enhancement agenda actively, leading the *New York Times* to profile it in a lengthy article as 'Singularity University, where you major in immortality'.

70 The Presidential Commission for the Study of Bioethical Issues (http://bioethics.gov), established by President Obama in 2009 and chaired by Professor Amy Gutmann. In August 2011 it announced that it would study the abuses of research ethics in Guatemala during the 1940s, when subjects were deliberately infected with sexually transmitted diseases (see Chapter 7).

10 key concepts

See text and boxes in appropriate chapter:

71 Justice (Chapter 1)
72 Property rights in the body (Chapter 2)
73 Exploitation and vulnerability (Chapters 1, 2 and 7)
74 Genetic patentability (Chapter 5)
75 Biological determinism (Chapter 4)
76 The genetic mystique/genetic exceptionalism (Chapter 4)
77 Killing and letting die (Chapter 6)
78 Autonomy (Chapter 1)
79 The genetic commons (Chapter 5)
80 The moral and legal status of the embryo or fetus (Chapter 6)

10 key thinkers

See text and boxes in appropriate chapter:

81 Aristotle (Chapter 1)
82 John Stuart Mill (Chapter 2)
83 Immanuel Kant (Chapter 2)
84 Richard Titmuss (Chapter 2)
85 Jeremy Bentham (Chapter 3)
86 James Boyle (Chapter 5)
87 John Locke (Chapter 5)
88 Karl Marx (Chapter 5)
89 Catherine Waldby and Melinda Cooper (Chapter 6)
90 Francis Bacon (Chapter 8)

10 key individuals and groups who have shaped the field, for better or worse

See text and boxes in appropriate chapter:

91 Hwang Woo Suk and Korean Womenlink (who revealed the truth about his research) (Chapters 1, 6 and 8)
92 The Seattle 'God Committee' (Chapter 1)
93 Tom Beauchamp and James Childress (Chapter 1)
94 The Tuskegee 'researchers' (Chapters 1 and 8)
95 Louise Brown, the first in vitro fertilization baby (Chapter 2)
96 Francis Collins and John Sulston, Human Genome Project (Chapter 4)
97 Richard Dawkins (Chapters 4 and 8)
98 Ananda Chakrabarty, the first researcher to patent a life form (Chapter 5)
99 James Thompson, Ian Wilmut and Shinya Yamanaka, stem cell researchers (Chapter 6)
100 Susan Reverby, the research ethics campaigner who uncovered the Guatemalan case (Chapter 7)

Notes

Chapter 1

1. Darnovsky, M., 'Moral questions of an altogether different kind: progressive politics in the biotech age', *Harvard Law and Policy Review* 2010; 4: 99–119.
2. Matas, D. and Kilgour, D., *Bloody Harvest: Revised Report into Allegations of Organ Harvesting from Falun Gong Prisoners in China* (Hamilton, Ontario: Seraphim Editions, 2009). Kilgour and Matas were nominated for the Nobel peace prize in 2010 for their work.
3. Dickenson, D. *Body Shopping: Converting Body Parts to Profit* (Oxford: Oneworld, 2009).
4. Pande, A., 'Commercial surrogacy in India: manufacturing a perfect mother-worker', *Signs: Journal of Women in Culture and Society* 2010; 35: 965–92. This practice was also documented in the 2011 documentary film *Made in India*, directed by Rebecca Haimowitz and Vaishali Sinha.
5. Franklin, S., 'Ethical biocapital', in Franklin, S. and Lock, M. (eds), *Remaking Life and Death: Towards an Anthropology of the Biosciences* (Santa Fe, NM: Society of American Research Press, 2003), p. 100.
6. Harris, J., 'Scientific research is a moral duty', *Journal of Medical Ethics* 2005; 31: 242–8, at p. 242.
7. Ibid.
8. For more detail, see Dickenson, *Body Shopping*; Spar, D.L., *The Baby Business: How Money, Science and Politics Drive the Commerce of Conception* (Cambridge, MA: Harvard Business School Press, 2006); and Cooper, M., *Life as Surplus: Biotechnics and Capitalism in the Neoliberal Era* (Seattle, WA: University of Washington Press, 2008).
9. Maranto, G., 'Ethical imaginaries', *Biopolitical Times*, 30 March 2011. Available at: www.biopoliticaltimes.org/article.php?id=5651 (accessed 4 April 2011).
10. Elliott, C., *White Coat, Black Hat: Adventures on the Dark Side of Medicine* (Boston: Beacon Press, 2010), p. 3 ff.
11. Rennie, S., 'Viewing research participation as a moral obligation: in whose interests?', *Hastings Center Report* 2011; 41: 40–7.
12. Baylis, F., 'For love or money? The saga of the Korean women who provided eggs for embryonic stem cell research', *Journal of Theoretical Medicine and Bioethics* 2009; 30: 385–96.

13. Kitzinger, J., 'Questioning the sci-fi alibi: a critique of how science-fiction fears are used to explain away public concerns about risk', *Journal of Risk Research* 2009; 13: 73–86.

14. Adapted from Dickenson, D., 'Unseen rise of body shopping', *The Sunday Times*, 20 April 2008.

15. Alexander, S., 'They decide who lives, who dies', *Life* 1960; 53: 102–25.

16. Hope, T., *Medical Ethics: A Very Short Introduction* (Oxford: Oxford University Press, 2004), p. 69.

17. Evans, J.H., 'A scholarly account of the growth of principlism', *Hastings Center Report* 2000; 30: 31–8.

18. Beauchamp, T. and Childress, J., *Principles of Biomedical Ethics*, 1st edn (New York: Oxford University Press, 1979) (5th edition 2009).

19. Dickenson, D., 'Cross-cultural issues in European bioethics', *Bioethics* 1999; 3: 249–55.

20. Senitulli, L., 'They came for sandalwood, now the b...s are after our genes!' Paper presented at the conference 'Research ethics, tikanga Maori/indigenous and protocols for working with communities,' Wellington, New Zealand, June 2004.

21. Sherwin, S., *No Longer Patient: Feminist Ethics and Health Care* (Philadelphia, PA: Temple University Press, 1993); McLeod, C. and Baylis, F., 'Feminists on the inalienability of human embryos', *Hypatia* 2006; 21: 1–24.

22. Pappworth, M., *Human Guinea Pigs: Experimentation on Man* (Boston: Beacon Press, 1967).

23. Elliott, *White Coat, Black Hat*, p. 170.

Chapter 2

1. Reported in 'Girls! Sell your eggs and enjoy the night life of Chennai,' *Bioedge*, 4 July 2009. Available at: www.bioedge.org (accessed 11 April 2011).

2. Waldby, C. and Cooper, M., 'The biopolitics of reproduction: post-Fordist biotechnology and women's clinical labour', *Australian Feminist Studies* 2008; 23: 57–73.

3. Jasanoff, S. *Designs on Nature: Science and Democracy in Europe and the United States* (Princeton, NJ: Princeton University Press, 2005).

4. Johnson, S., 'Ethics of reproductive tourism questioned', *Blog Bioethics Net*, 20 May 2010.

5. Robert Hintze, quoted in 'It's a moral imperative', *Bioedge*, 25 June 2010. Available at: www.bioedge.org (accessed 12 April 2011).

6. Beth Goodman, quoted in Nosheen, H. and Schillmann, K., 'The most wanted surrogates in the world', *Biopolitical Times*, 20 October 2010. Available at: www.geneticsandsociety.org/article.php?id=5421 (accessed 29 August 2011).

7. *On Liberty* (1859), ch. 1.

8. HFEA, *SEED Report: A Report on the HFEA's Review of Sperm, Egg and Embryo Donation in the United Kingdom* (London: HFEA, 2006), section 4.3.

9. Marquadt, E., Glenn, N.D. and Clark, K., *My Daddy's Name is Donor* (New York: Center for American Values, 2010).

10. Marquadt, E., 'Is the glass half full or half empty? Debating the research on donor offspring: A reply to Blyth and Kramer's critique of *My Daddy's Name is Donor*', *Bionews*, 9 August 2010. Available at: www.bionews.org.uk (accessed 12 April 2011).

11. Anonymous comment on Vickers, H., 'Sperm and egg donors should be paid more, experts claim', *Bionews*, 25 October 2010. Available at: www.bionews.org.uk (accessed 25 October 2010).

12. Titmuss, R., *The Gift Relationship* (London: Allen and Unwin, 1970).

13. www.eggdonation.com, a California agency which promises customers that 'we will personally guide you in finding your donor angel'. It also assures potential egg sellers that 'we will ensure that your journey is safe and gratifying and we will reward you for your gesture with commemorating gifts and the highest level of compensation. We will treat you like the angel you are' (accessed 11 April 2011).

14. Johnson *v* Calvert (5 Cal 4th 84) established in 1993 that the woman who first expresses her intention to raise the child has custody, although neither the court nor the California legislature has stated that surrogacy and egg sale are legal. In the UK two apparently contradictory cases were decided early in 2011: one, *Re L*, allowing a British couple who had commissioned a US contract pregnancy to retain custody of the child, as being in the child's best interests, but the other upholding a birth mother's right to refuse to hand over the baby on the grounds that she had developed a bond with the child. Both were actually decided in the name of the child's best interests and so are not as contradictory as they seem.

15. Thernstrom, M., 'My futuristic insta-family,' *New York Times Magazine*, 2 January 2011, pp. 34–5. Thernstrom refers to the woman who sold her the eggs used to create the two embryos as her 'Fairy Goddonor' (original capitals).

16. Quoted in *Bionews* 580, 18 October 2010. Available at: www. bionews.org (accessed 11 April 2011).
17. Goodwin, M.B. (ed.), *Baby Markets: Money and the New Politics of Creating Families* (Cambridge: Cambridge University Press, 2010).
18. Spar, D., *The Baby Business: How Money, Science and Politics Drive the Commerce of Conception* (Cambridge, MA: Harvard University Press, 2006).
19. In Goodwin (ed.), *Baby Markets*, pp. 23–40, at p. 23.
20. Quoted in *Bioedge*, 12 June 2009. Available at: www.bioedge.org (accessed 11 April 2009).
21. A surrogate mother interviewed by Abigail Howarth in her 2007 article, 'Surrogate mothers – womb for rent', *Marie Claire*, 29 July.
22. Widdows, H., 'Ethics and global governance: the poverty of choice', Professorial Inaugural Lecture, University of Birmingham, 24 March 2011.
23. Quoted in Pande, A., 'Commercial surrogacy in India: manufacturing a perfect mother-worker', *Signs: Journal of Women in Culture and Society* 2010; 35: 965–92.
24. Pande, op. cit.
25. Quoted in *Bioedge*, 22 May 2010. Available at: www.bioedge.org (accessed 11 April 2009).
26. Satz, D., *Why Some Things Should Not Be For Sale: The Moral Limits of Markets* (Oxford: Oxford University Press, 2011).
27. Honore, A.M., 'Ownership', in A.G. Guest (ed.), *Oxford Essays in Jurisprudence* (Oxford: Oxford University Press, 1961); Dickenson, D., *Property in the Body: Feminist Perspectives* (Cambridge: Cambridge University Press, 2007), p. 13.
28. *The Visible and the Invisible* (Evanston, IL: Northwestern University Press, 1968), p. 37
29. Merchant, J. 'Le pouvoir de procréer par assistance médicale; la vision de la société nord-américaine', in *L'embryon, le foetus, l'enfant: assistance médicale à la procréation (AMP) et lois de bioéthique* (Paris, Edition ESKA, 2009).
30. Dickenson, D., *Property, Women and Politics* (Cambridge: Polity Press, 1997); Dickenson, D., 'Property and women's alienation from their own reproductive labour', *Bioethics* 2001; 15: 205–17.
31. Kramer, W., Schneider, J. and Schultz, N., 'US oocyte donors: a retrospective study of medical and social issues', *Human Reproduction* 2009, 10.193/humrep/dep309 (accessed 12 April 2011).
32. Gurmankin, A.D., 'Risk information provided to prospective oocyte donors in a preliminary phone call', *American Journal of Bioethics* 2002; 1: 4.

33. Levine, A.D., 'Self-regulation, compensation and the ethical recruitment of egg donors', *Hastings Center Report* 2010; 40: 25–36. In a survey of advertisements in 83 US college newspapers, Levine found that nearly a quarter of advertisements broke the American Society for Reproductive Medicine (ASRM) guidelines for egg sale of $5,000 (exceptionally $10,000). Of the advertisements violating ASRM guidelines, many offered $20,000, several offered $35,000 and one was as high as $50,000.

34. Coeytaux, F., Darnovsky, M. and Fogel, S. 'Editorial: Assisted reproduction and choice in the biotech age: recommendations for a way forward,' *Contraception* 2011; 1 Jan.

35. 'A robust, particularist assessment of medical tourism', *Developing World Bioethics* 2010; 11: 16–29.

36. Quoted in Pet, D., 'India moves toward regulation of assisted reproduction and surrogacy', *Biopolitical Times*, 10 February 2010. Available at: www.biopoliticaltimes.org (accessed 12 April 2011).

37. Ikemoto, L., 'Eggs as capital: human egg procurement in the fertility industry and the stem cell research enterprise', *Signs* 2009; 34: 763–81, at pp. 768 and 779.

38. Argentina, also technologically advanced and 'light touch' in regulation, has a similar thriving market in egg sales. See Smith, E., Behrman, J., Martin, C. and Williams-Jones, B., 'Reproductive tourism in Argentina: clinic accreditation and its implications for consumers, health professionals and policy makers', *Developing World Bioethics* 2010; 10: 59–69.

39. Humbyrd, C., 'Fair trade international surrogacy', *Developing World Bioethics* 2009; 9: 111–19.

40. 'Justice and the market domain', in J.W. Chapman and J.R. Pennock (eds), *Nomos XXII: Property* (New York: NYU Press, 1989), p.175.

Chapter 3

1. Levitt, M. and O'Neill, F.K., 'Making humans better and making better humans', *Genomics, Society and Policy* 2011; 6: 1–14.

2. Buchanan, A., *Beyond Humanity? The Ethics of Biomedical Enhancement* (Oxford: Oxford University Press, 2011), p. 23.

3. Rose, N., *The Politics of Life Itself: Biomedicine, Power and Subjectivity in the Twenty-First Century* (Princeton: Princeton University Press, 2007), p. 17.

4. Habermas, J., *The Future of Human Nature* (Cambridge: Polity, 2003), quoted in Buchanan, *Beyond Humanity*, p. 5.

5. Agar, N., *Liberal Eugenics: In Defence of Human Enhancement* (Oxford: Blackwell, 2005).

6. Dickenson, D. *Body Shopping: Converting Body Parts to Profit* (Oxford: Oneworld, 2009), p. 3.

7. Schneider, S.W., 'Jewish women's eggs: a hot commodity in the IVF marketplace', *Lilith* 2001; 26: 22; Levine, A.D., 'Self-regulation, compensation and the ethical recruitment of oocyte donors', *Hastings Center Report* 2010; 40: 25–36.

8. Parker, M., 'The best possible child', *Journal of Medical Ethics* 2007; 33: 279–83.

9. Lo, Y.M.D., et al., 'Maternal plasma DNA sequencing reveals the genetic and mutational profile of the fetus', *Science and Translational Medicine* 2010; 2: 61.

10. Darnovsky, M., 'One step closer to designer babies', *Biopolitical Times*, 22 April 2011. Available at: www.geneticsandsociety.org/article.php?id=5687 (accessed 29 August 2011).

11. Human Genetics Commission, *Increasing Options, Informing Choice: A Report on Preconception Genetic Screening* (London: Human Genetics Commission, 2011).

12. Allahbadia, G.N., 'The 50 million missing women', *Journal of Assisted Reproduction and Genetics* 2002; 19: 411–16.

13. Thiele, A.T. and Their, B., 'Towards an ethical policy for the prevention of fetal sex selection in Canada', *Journal of Obstetrics and Gynecology Canada* 2010; 32: 54–7.

14. de Saille, S., 'Is sex selection illegal and immoral?', *Bionews*, 28 March 2011.

15. Savulescu, J., 'Procreative beneficence: why we should select the best children', *Bioethics* 2001; 15: 413–26; 'Deaf lesbians, designer disability and the future of medicine', *British Medical Journal* 2002; 325: 771–3.

16. Fried, C., *Right and Wrong* (Cambridge, MA: Harvard University Press, 1978), p. 13.

17 Savulescu, J., 'Personal choice: letter from a doctor as a dad', in K. W. M. Fulford, D. L. Dickenson and T. H. Murray (eds), *Healthcare Ethics and Human Values* (Oxford: Blackwell, 2002), pp. 109–10.

18. Richards, J., 'But didn't you have the tests?', in Fulford, Dickenson and Murray (eds), *Healthcare Ethics*, pp. 232–5, at pp. 234–5.

19. Parker, 'The best possible child', p. 281.

20. le Doeuff, M., *Le sexe du savoir* [*The Gender of Knowledge*] (Paris: Champs Flammarion, 2000), translation mine.

21. Harris, J., *Enhancing Evolution* (Princeton, NJ: Princeton University Press, 2007), p. 9.

22. Huxley, A., *Brave New World* (New York: HarperCollins, 1994), p. 217.

23. Buchanan, *Beyond Humanity*, p. 221.
24. Annas, G., 'Cell division', *Boston Globe*, 21 April 2005, cited in Buchanan, *Beyond Humanity*, p. 225.
25. Buchanan, *Beyond Humanity*, p. 226.
26. de Andrade, N.N.G., 'Human genetic manipulation and the right to identity: the contradictions of human rights law in regulating the human genome', *Scripted* 2010; 7 (December).
27. Sandel, M., 'The case against perfection: what's wrong with designer children, bionic athletes and genetic engineering', *Atlantic Monthly* 2004 (April). Available at: www.theatlantic.com/past/docs/issues/2004/sandel.htm (accessed 4 May 2011).

Chapter 4

1. Nelkin, D. and Lindee, M.S., *The DNA Mystique: The Gene as Cultural Icon* (New York: WH Freeman and Company, 1995), pp. 41–2.
2. Paxman, R., 'Study suggests gene linked to credit card debt', *Bionews*, 10 May 2010. Available at: www.bionews.org.uk/page_59621.asp (accessed 10 May 2011). This study of 2,500 18- to 26-year-olds, which had not appeared in a peer-reviewed journal at the time it was publicly announced, posited a link between the *MAOA* (monoamine oxidase A) gene and impulsive behaviours, including running up credit card debt.
3. Knafo, A., et al., 'Individual differences in allocation of funds in the dictator game associated with length of the arginine vasopressin 1a receptor RS3 promoter region and correlation between RS3 length and hippocampal mRNA', *Genes, Brain and Behaviour* 2008; 7: 266–75.
4. Fowler, J.H. and Dawes, C.T., 'Two genes predict voter turnout', *Journal of Politics* 2007; 70: 579–94.
5. Medved, M. (2008), Respecting – and recognizing – American D.N.A., Available at: www.townhall.com, (accessed 12 May 2011).
6. Chadwick, R., 'Are genes us? Gene therapy and personal identity', in G.K. Becker (ed.), *The Moral Status of Persons* (Amsterdam: Rodopi, 2000), pp. 183–94.
7. This is, of course, an oversimplification: the Navajo are more recent immigrants to the American Southwest than the Pueblo peoples, for example, while the Pueblo themselves vacated some sites, such as Frijoles Canyon at Bandelier National Monument northwest of Santa Fe, and migrated to others.
8. 'Free will an illusion, says noted US biologist', *Bioedge*, 15 February 2010.

9. Dawkins, R., *The Selfish Gene* (Oxford: Oxford University Press, 1976).

10. Many of these definitions appear in the *Bionews* glossary at www.bionews.org.uk.

11. Fatimathas, L., 'Happy disposition? New study claims it could be in your genes', *Bionews*, 9 May 2011. Available at: www.bionews.org.uk/page_94153 (accessed 12 May 2011).

12. Amos, C.I., Spitz, M.I. and Cinciripini, P., 'Chipping away at the genetics of smoking behaviour', *Nature Reviews Genetics* 2010; 42: 366–8.

13. Collins, F., *The Language of Life: DNA and the Revolution in Personalized Medicine* (New York: HarperCollins, 2010).

14. Sulston, J. (with Ferry, G.), *The Common Thread: Science, Ethics, Politics and the Human Genome* (London: Corgi, 2003); Collins, *The Language of Life*.

15. Gee, H., quoted in 'Editorial: Best is yet to come', Nature 2011; 470; 140, 9 February, doi: 10.1038/470140a.

16. GeneWatch, *History of the Human Genome* (Buxton: Genewatch UK, June 2010).

17. Wade, N., 'A decade later, genetic map yields few cures', *New York Times*, 12 June 2010.

18. Paynter, N.P., et al., 'Cardiovascular risk disease prediction with and without knowledge of genetic variation at chromosome 9.p21.3', *Annals of Internal Medicine* 2009; 150: 65–72.

19. Brody, H., *The Future of Bioethics* (New York: Oxford University Press, 2009), p. 9.

20. Quoted in Pearson, H., 'One gene, twenty years', *Nature* 2009, 460: 164–9.

21. Heard, E., et al., 'Ten years of genetics and genomics: what have we learned and where are we heading?', *Nature Reviews Genetics* 2010; 11: 723–33, at p. 723.

22. Monk, M., 'The new epigenetics', *Bionews*, 9 November 2009. For a further discussion of recent findings on the interaction between environment and the functioning of genes, see Carey, B., 'Genes as mirrors of life experiences', *New York Times*, 8 November 2010.

23. For a fuller discussion of the Tongan case, see my two books *Body Shopping* (p. 104 ff.) and *Property in the Body* (p. 162 ff.)

24. Senituli, L., 'They came for sandalwood, now the b...s are after our genes!' Paper presented at the conference 'Research ethics, Tikanga Maori/indigenous and protocols for working with communities', Wellington, New Zealand, 10–12 June 2004.

25. Mead, H.M., *Tikanga Maori: Living by Maori Values* (Wellington, NZ: Huia Publishers, 2003).

26. For case examples involving this dilemma, see Dickenson, D., Huxtable, R. and Parker, M., *The Cambridge Medical Ethics Workbook*, 2nd edn (Cambridge: Cambridge University Press, 2010), Chapter 3, 'Genetics: information, access and ownership'.

Chapter 5

1. Jensen, K. and Murray, F., 'International patenting: the landscape of the human genome', *Science* 2005; 310: 239–40.

2. *Diamond v Chakrabarty* (1980) 447 US 303.

3. Knowles, L. B., 'Of mice and men: patenting the onco-mouse', *Hastings Center Report* 2003; 33: 6–7.

4. In the *Howard Florey/Relaxin* case (European Patent Office Reports, 1995, p. 541), the German Green Party argued that a patent on a synthetic form of relaxin, a hormone secreted in pregnant women, amounted to slavery because it involved the dismemberment of female tissue and its sale to profit-making companies. This argument was rejected on the grounds that the women who provided the genetic material had given informed consent, and that no individual woman's tissue was being patented, because that's not how the synthetic version was produced.

5. Andrews, L.B., 'Genes and patent policy: rethinking intellectual property rights', *Nature Reviews Genetics* 2002; 3: 803–8.

6. Goldman, B., 'HER2: the patent "genee" is out of the bottle', *Journal of the Canadian Medical Association* 2007; 176: 1443–4; Barrett, A., et al., 'How much will Herceptin really cost?', *BMJ* 2006; 333:1118.

7. Cooper, M., *Life as Surplus: Biotechnology and Capitalism in the Neo-Liberal Era* (Seattle: University of Washington Press, 2008).

8. *Association for Molecular Pathology et al. v US Patent and Trade Office et al.*, 669 F Supp 2d 265 (29 March 2010).

9. Eisenberg, R.S., 'How can you patent genes?', *American Journal of Bioethics* 2002; 2: 3–11, at p. 4.

10. For further development of this distinction, see Waldron, J., *The Right to Private Property* (Oxford: Clarendon Press, 1988).

11. Dickenson, D., 'The lady vanishes: what's missing from the stem cell debate', *Journal of Bioethical Inquiry* 2006; 3: 43–54.

12. Brief for the United States as *amicus curiae* ['friend of the court', an advisory brief] in support of neither party, filed 29 October 2010, no. 2010-1406, in: *Association for Molecular Pathology et al. v US Patent and Trade Office and Myriad Genetics Inc.*, at p. 10. The

amicus curiae brief is somewhat ambivalent: it does seem to allow a get-out clause for cDNA, a copied form, to be patented.

13. For example, Article 1: 'The human genome underlies the fundamental unity of all members of the human family, as well as the recognition of their inherent dignity and diversity. In a symbolic sense, it is the heritage of humanity.'

14. Boyle, J., 'The second enclosure movement and the construction of the public domain', *Law and Contemporary Problems* 2003; 66: 33–74, at p. 37.

15. Hardin, G., 'The tragedy of the commons', *Science* 1968; 162: 1243 ff.

16. Gunn, N.M., *Butcher's Broom* (Edinburgh: Polygon, 2006), pp. 231–2.

17. Dickenson, D., *Body Shopping: Converting Body Parts to Profit* (Oxford: Oneworld, 2009), p. 159.

Chapter 6

1. Brown, E., 'CIRM funds Geron Corp. spinal cord injury trial,' *Los Angeles Times*, 4 May 2011.

2. Girgis, S., 'The scientists knew they were lying?', *Public Discourse*, 13 April 2011. Available at: www.thepublicdiscourse.com/2011/04/2490 (accessed 27 May 2011).

3 Darnovsky, M., 'Stem cell politics and progressive values', *Biopolitical Times*, 15 June 2006. Available at: www.geneticsandsociety.rsvp1.com/article.php?id=1951&mgh=http%3A%2F%2Fwww.geneticsandsociety.org&mgf=1 (accessed 27 May 2011).

4. Jackson, E., 'Fraudulent stem cell research and respect for the embryo', *Biosciences* 2006; 1: 349–56. This seems a particularly odd position to take because the SCNT technique creates clones of the person who donated the somatic cell used with the enucleated egg. Reproductive cloning is illegal in almost all jurisdictions, although there is no specific federal law banning it in the United States.

5. Thompson, J.A., et al., 'Embryonic stem cell lines derived from human blastocysts', *Science* 1998; 282: 1145–7.

6. Takahashi, K., et al., 'Induction of pluripotent stem cells from adult fibroblasts by defined factors', *Cell* 2007; 131: 861–72.

7. Zhao, T., et al. 'Immunogenicity of induced pluripotent stem cells', *Nature*, 13 May 2011. Available at: www.nature.com (accessed 27 May 2011).

8. Leeb, C. et al., 'Promising new sources for pluripotent stem cells', *Stem Cell Rev and Rep* 2010; 6: 15–26.

9. Macchiarini, P., et al., 'Clinical transplantation of a tissue-engineered airway,' *Lancet* 2011; 372: 2023–30.

10. Laughlin, M.J., et al., 'Hematopoietic engraftment and survival in adult recipients of umbilical-cord blood from unrelated donors', *N Engl J Med* 2001; 344: 1815–22.

11. Ballen, K., 'Challenges in umbilical cord blood stem cell banking for stem cell reviews and reports', *Stem Cell Rev and Rep* 2010; 6: 8–14.

12. Brown, N., Machin, L. and McCleod, D., 'The immunitary bioeconomy: the economisation of life in the international cord blood market', *Social Science and Medicine* 2011; 30: 1–8, doi:10.1016/jsocscimed.2011.01.024.

13. (1978) 10 Pa.D. and C. 3d 90.

14. Dickenson, D., 'Good science and good ethics: why we should discourage payment for eggs in stem cell research', *Nature Reviews Genetics* 2009; 10: 743; Hyun, I., 'Stem cells from skin cells: the ethical questions', *Hastings Center Report* 2008; 38: 20–2.

15. Waldby, C. and Cooper, M., 'From reproductive work to regenerative labour: the female body and the stem cell industries', *Feminist Theory* 2010; 11: 3–22, emphasis added.

16. Ballen 'Challenges in umbilical cord blood', p. 11.

17. The RCOG received legal advice that umbilical cord blood is legally the mother's and not the baby's, although few commentators seem aware of that.

18. Ediezen, L.C., 'NHS maternal units should not encourage the private banking of umbilical cord blood', *BMJ* 2006; 333: 801–4.

19. Thornlet, I., et al., 'Private cord blood banking: experiences and views of pediatric hematopoietic cell transplantation physicians', *Pediatrics* 2009; 123: 1011–17.

20. MacKenna, R., 'Umbilical cord blood banking not worth the cost, study shows', *Bionews*, 28 September 2009.

21. Sleeboom-Faulkner, M. and Patna, P.K., 'The bioethics vacuum: national policies on human embryonic stem cell research in India and China', *Journal of International Biotechnology Law* 2008; 5: 221–34.

Chapter 7

1. Susan Reverby, quoted in Kasdon, L., 'A dark study comes to light', *Wellesley College Alumnae Magazine* 2011(Winter), pp. 37–9, at p. 38.

2. A recent review by the Associated Press of medical journals and press clippings found more than 40 US studies involving prisoners and mental patients (cited in Mike Stobbe, 'Past medical testing on humans revealed,' *Washington Post*, 27 February 2011).

3. Jones, J.H., *Red Blood: The Tuskegee Syphilis Experiment* (Glencoe, IL: Free Press, 1993); Reverby, S.M. (ed.), *Tuskegee's Truths: Rethinking the Tuskegee Syphilis Study* (Chapel Hill: University of North Carolina Press, 2000); Reverby, S.M., *Examining Tuskegee: The Infamous Syphilis Study and its Legacy* (Chapel Hill: University of North Carolina Press, 2008).
4. Art Caplan, interviewed in Stobbe, 'Past medical testing in humans revealed'.
5. Quoted in Kaiser, J., 'US bioethics panel to review clinical trials around the world', *Science Insider*, 1 March 2011. Available at: http://news.science.mag.org/scienceinsider/2011/03.us-bioethics-panel-to-review.html (accessed 10 June 2011).
6. From: *Trials of war criminals before the Nuremberg military tribunals under Control Council law no. 10: volume 2, Nuremberg October 1946–April 1949* (Washington DC: US Government Printing Office, 1949), pp. 181–2. See also: Annas, G.J. and Goodin, M.A. (eds), *The Nazi Doctors and the Nuremberg Code: Human Rights in Human Experimentation* (Oxford: Oxford University Press, 1992).
7. Quoted in Stobbe, 'Past medical testing in humans revealed'.
8. McFarlane, S., 'President's bioethics commission continues review of Guatemalan syphilis experiments', *Biopolitical Times*, 26 May 2011.
9. Dr John Cutler, from the archives discovered by Susan Reverby, quoted in McFarlane, 'President's bioethics commission'.
10. US National Institutes of Health and Centers for Disease Prevention and Control (1997) 'A defense of HIV trials in the developing world'. Available at: www.nih.gov/%20news/mathiv/mathiv.htm (accessed 10 June 2011). For the opposing view, see Angell, M., 'Ethical imperialism? Ethics in international collaborative clinical research', *N Engl J Med* 1998; 319: 1081–3; Lurie, P. and Wolfe, S.M., 'Unethical trials of interventions to reduce perinatal transmission of the human immunodeficiency virus in developing countries', *N Engl J Med* 1997; 337: 853–6.
11. Quoted in Bentley, S. and Smith, D., 'As doctors fought to save lives, Pfizer flew in drug trial team,' *Guardian*, 10 December 2010.
12. UK Academy of Medical Sciences, *A New Pathway for the Regulation and Governance of Health Research* (London: Academy of Medical Sciences, 2011).
13. Elliott, C., *White Coat, Black Hat: Adventures on the Dark Side of Medicine* (Boston: Beacon Press, 2010), p. 20.

14. Steinbrook, R., 'Compensation for injured trial subjects', *N Engl J Med* 2006; 354: 1871–3. Admittedly, clinical trial subjects are still in a better position than egg 'donors', who get neither follow-up care nor even monitoring. See Kramer, W., Schneider, J. and Schultz, N., 'US oocyte donors: a retrospective study of medical and social issues,' *Human Reproduction Online* 2009; doi: 10.193/humrep/dep309.
15. Perry, S., 'Too many clinical trials still exploit the poor and other vulnerable people, says U of M bioethics professor', *Minnesota Post*, 14 October 2010.
16. Steinbrook, R. (2005) 'Gag clauses in clinical trials agreements,' *N Engl J Med* 2005; 352: 2180–2.
17. Mello, M. M. and Wolf, L. E. 'The Havasupai Indian tribe case – lessons for research involving stored biologic samples', *N Engl J Med*, 9 June 2010, doi: 10.1056/NEJMp1005203.
18. For a discussion of a 2007 case involving a legal challenge to transfer of stored prostate tissue samples against the donors' wishes (the Catalona case), see my *Body Shopping*, Chapter 6.
19. McHale, J., 'Accountability, governance and biobanks: the ethics committee as guardian or "toothless tiger"?'. Paper given at European Commission TissEU project workshop, Birmingham, 4 June 2010.

Chapter 8

1. Lopatin, P., 'What scientists believe', *New Atlantis* 2010 (Fall). Available at: www.thenewatlantis.com/publications/what-scientists-believe (accessed 8 June 2011).
2. Dawkins, R., *The God Delusion* (London: Bantam Press, 2006), title of Chapter 8.
3. Dawkins, *The God Delusion*, p. 284.
4. Dawkins, *The God Delusion*, p. 281.
5. Cooper, M., 'The unborn born again – neo-imperialism, the evangelical right and the culture of life', *Postmodern Culture* 2006; 17–38.
6. Fox, R.C. and Swazey, J.P., *Observing Bioethics* (New York: Oxford University Press, 2008).
7. For the viewpoint of the firm, Sciona, see Finegold, D.L., et al., *Bioindustry Ethics* (Amsterdam: Elsevier, 2005).
8. Quoted in Frankenberry, N., *The Faith of Scientists: In Their Own Words* (Princeton, NJ: Princeton University Press, 2010).
9. Cited in Ecklund, E. H., *Science vs. Religion: What Scientists Really Think* (New York: Oxford University Press, 2010).

10. Burnell, J. B., 'Faith in the universe', lecture given at Oxford Friends Meeting, 26 May 2011.

11. Quoted in Evans, J.H., 'Science, bioethics and religion', in Harrison, P. (ed.), *Science and Religion* (Cambridge: Cambridge University Press, 2010), pp. 207–25, at p. 210.

12. Ecklund, *Science vs. Religion.*

13. Bacon, F. (1620) *On the Advancement of Learning* (London: Cassell and Company, 1893), section VI, paragraph 14.

14. Bacon, *On the Advancement of Learning*, section I, paragraph 3.

15. Bacon, *On the Advancement of Learning*, section V, paragraph 11, emphasis added.

16. Nelkin, D., 'Is bioethics for sale? The dilemmas of conflict of interest', *The Tocqueville Review* 2003; 24: 45–60.

17. Evans, 'Science, bioethics and religion', p. 218.

18. Evans, 'Science, bioethics and religion', pp. 218–19, citing Rosenberg, C. E., 'Meanings, policies and medicine: on the bioethical enterprise and history', *Daedalus* 1999; 128: 27–46, at p. 38.

19. In addition to the many references in other chapters, see also: McLeod, C. and Baylis, F., 'For dignity or money: feminists on the commodification of women's reproductive labour', in Steinbock, B. (ed.), *The Oxford Handbook of Bioethics* (New York: Oxford University Press, 2005), pp. 258–83; Parry, B., *Trading the Genome: Investigating the Commodification of Bio-Information* (New York: Columbia University Press, 2004); Widdows, H., 'Persons and their parts: new reproductive technologies and risk of commodification', *Health Care Analysis* 2009; 17: 36–46; and the essays in *Altruism's Limits*, ed. Michele Goodwin (New York: Cambridge University Press, forthcoming in 2012).

20. Klein, N., *No Logo: Taking Aim at the Brand Bullies* (Toronto: Knopf, 2000), and *The Shock Doctrine: The Rise of Disaster Capitalism* (Harmondswowrth: Penguin, 2007). Similar popular critiques include Reich, R., *Supercapitalism: The Battle for Democracy in an Age of Big Business* (New York: Knopf, 2007), and Elliott, L. and Atkinson, D., *The Gods that Failed: How Blind Faith in Markets Has Cost Us Our Future* (London: Bodley Head, 2008).

21. Email from Richard Haynes, co-convenor of the Tarrytown meetings, 6 December 2010.

Index

ALL THAT MATTERS: BIOETHICS